William Fairbairn

The Principles of Mechanism and Machinery of Transmission

Comprising the principles of mechanism, wheels and pulleys, strength and proportions of shafts, couplings for shafts, and engaging and disengaging gear

William Fairbairn

The Principles of Mechanism and Machinery of Transmission
Comprising the principles of mechanism, wheels and pulleys, strength and proportions of shafts, couplings for shafts, and engaging and disengaging gear

ISBN/EAN: 9783337164195

Printed in Europe, USA, Canada, Australia, Japan

Cover: Foto ©berggeist007 / pixelio.de

More available books at **www.hansebooks.com**

THE

PRINCIPLES OF MECHANISM

AND

MACHINERY OF TRANSMISSION.

COMPRISING

THE PRINCIPLES OF MECHANISM, WHEELS AND PULLEYS,
STRENGTH AND PROPORTIONS OF SHAFTS, COUPLINGS FOR
SHAFTS, AND ENGAGING AND DISENGAGING GEAR.

BY

WILLIAM FAIRBAIRN, ESQ., C.E.
LL. D. F.R.S. F.G.S.

CORRESPONDING MEMBER OF THE NATIONAL INSTITUTE OF FRANCE, AND OF THE
ROYAL ACADEMY OF TURIN; CHEVALIER OF THE
LEGION OF HONOUR, ETC., ETC.

PHILADELPHIA:
HENRY CAREY BAIRD,
INDUSTRIAL PUBLISHER,
No. 406 WALNUT STREET.
1869.

PUBLISHER'S PREFACE.

This work, now offered to the public, on the "Principles of Mechanism and Transmission," is taken from the work of the distinguished author on "Mills and Millwork," which is large and expensive, and contains many details of Millwork not entirely adapted to American practice. The great principles here given lie at the very foundation of the art of transmitting mechanical power, and must prove of inestimable value to American mill-owners, mechanics, and operatives.

In the present volume Mr. Fairbairn gives the results of his very successful practice as a Millwright and Engineer, during a period of half a century—a period which has contributed more than any previous one to the manufacturing industry of the world.

And as there is probably no department of practical science so generally useful, or perhaps so little studied, as the machinery of transmission, the American publisher has great pleasure in placing within reach of the intelligent class for which it is intended, in a compact and comparatively cheap form, the author's rich and valuable experience on this important subject.

CONTENTS.

CHAPTER I.

THE PRINCIPLES OF MECHANISM.

GENERAL VIEWS, LINK-WORK, WRAPPING CONNECTORS WHEEL-WORK, SLIDING CONTACT :—

	PAGE
General Views Relating to Machines	13
The parts of a Machine	18
Elementary Forms of Mechanism	27
Link-work	22

ELEMENTARY FORMS OF MECHANISM:—

To construct Watt's parallel motion	31
To multiply Oscillations by means of Link-work	34
To produce a Velocity which shall be rapidly retarded by means of Link-work	36
To produce a reciprocating intermittent Motion by means of Link-work	37
The Ratchet-wheel and Detent	39
Intermittent motion produced by Link-work connected with a Ratchet-wheel	39
Wrapping Connectors	40
Speed Pulleys	44
Guide Pulleys	47
To prevent Wrapping Connectors from Slipping	48
System of Pulleys	51
To produce a varying velocity ratio by means of Wrapping Connectors	54

CONTENTS.

	PAGE
Wheel-work producing Motion by rolling Contact..	56
Idle Wheels...................................	63
Annular Wheels—Concentric Wheels.............	64
Wheel-work when the axes are not parallel to each other..	64
Face-wheel and Lantern—Crown-wheels..........	65
To construct Bevel-wheels or Bevel-gear when the axes are in the same plane.....................	66
To construct Bevel-gear when the axes are not in the same plane.................................	68
Variable motions produced by Wheel-work having rolling contact................................	69
Intermittent and reciprocating motions produced by Wheel-work having rolling contact.............	71
The Wedge and Movable Inclined Plane..........	74
Sliding Pieces producing motion by sliding contact.	74
The Eccentric Wheel...........................	75
Cambs, Wipers, and Tappets....................	76
To find the curve forming the groove of a Camb, so that the velocity ratio of the rod and axes of the Camb may be constant.........................	77
The Swash Plate...............................	80
Construction of Screws.........................	82
The Solid Screw and Nut........................	85
The Common Press.............................	86
The Compound Screw..........................	88
The Endless Screw.............................	89
The Differential Screw.........................	90
The Archimedian Screw Creeper................	91
Mechanism for Cutting Screws..................	92
To produce a changing reciprocating rectilinear motion by a combination of the Camb and Screw.	94
To produce a boring motion by a combination of the Screw and Toothed Wheels:................	95

CHAPTER II.

ON MACHINERY OF TRANSMISSION.

ON WHEELS AND PULLIES :—

	PAGE
Wrapping Connections	99
Where employed	100
Advantages and Disadvantages of	101
Material employed in the Construction of	101
Strength of	102
Table of approximate Widths of Leather Straps, in Inches, necessary to transmit any required Number of Horses' Power	103

TOOTHED WHEELS :—

Introduction of	104
Construction of Mortise Wheels	105
Smeaton's Introduction of Cast-iron as a Material for Spur Wheels	107
Rennie's use of Cast-iron in all the details of Millwork, as exemplified in the Construction of the Albion Mills	107
True Principle of Construction	108
Tooth-cutting Machine	112

SPUR GEARING :—

Definitions	114

PITCH OF WHEELS :—

Rules for finding the Pitch and Diameter of Wheels	117
Table of Constants for Wheel-work	118
Rules for finding the Pitch, Diameter, and Number of Teeth	119
Professor Willis's Method of graduating the sizes of Wheels	121
Table showing the relation of Pitch, Diameter, and Number of Teeth	122, 123

CONTENTS.

PAGE

TEETH OF WHEELS :—
  The Principles which determine the proper Form.. 124
  Formation of Epicycloidal and Hypocycloidal Curves 125
  Construction of Epicycloidal Teeth 129
  Construction of Involute Teeth.................. 135
  Professor Willis's Method of striking the Teeth of Wheels.. 140
  Odontograph.....................................142
  General Form and Proportions of Teeth of Wheels. 145
  Table of Proportions of Teeth of Wheels for average Practice 154
  Table giving the Proportions of the Teeth of Wheels in Inches and Thirty-seconds of an Inch 156

BEVEL WHEELS :—
  Examination of the Curves....................... 157
  Formation and Form of Teeth.................... 159

SKEW BEVELS :—
  Definitions and Method of setting out the Teeth... 160

THE WORM AND WHEEL:
  Description of Construction..................... 163

STRENGTH OF THE TEETH IN WHEELS :—
  Rules to be observed in Calculations............. 165
  Line of greatest Strain.......................... 167
  Table of Thickness, Breadth, and Pitch of Teeth of Wheels.. 168
  Table of Relation of Horses' Power transmitted and Velocity at the Pitch Circle to Pressure on Teeth 172
  Table showing the Pitch and Thickness of Teeth to transmit a given Number of Horses' Power at different Velocities............................ 173

CONTENTS.

PAGE

Table showing the Breadth of Teeth required to transmit different Amounts of Force at a uniform Pressure of 400 lbs. per inch.................. 174

CHAPTER III.

ON THE STRENGTH AND PROPORTIONS OF SHAFTS:—
The Factory System necessitates the use of long Ranges of Shafts............................. 175

DIVISION I.:—
The Material of which Shafting is constructed.... 177

DIVISION II. TRANSVERSE STRAIN:—
Resistance to Rupture.......................... 179
Rules for the Strength of Shafts................. 183
Table of Resistance to Flexure. Weights producing a Deflection of $\frac{1}{1500}$ of the Length in Cast-iron Cylindrical Shafts....................... 187
Table of Resistance to Flexure. Weights producing a Deflection of $\frac{1}{1500}$th of the length in Wrought-iron Cylindrical Shafts............... 188
Table of Deflection of Cast-iron Cylindrical Shafts, arising from the Weight of the Shaft........... 189
Table of Deflection of Wrought-iron Cylindrical Shafts, arising from the Weight of the Shaft.... 190

DIVISION III. TORSION:—
Coulomb's Deductions and Formula.............. 191
Bevan's Values of Modulus of Torsion........... 192
Wertheim's Formulæ for Cylindrical Bodies....... 193
Résumé of Experiments on Cylinders of Circular Section...................................... 196
Résumé of Experiments on the Torsion of Hollow Cylinders of Copper......................... 197
Résumé of Experiments on the Torsion of Elliptical Bars...................................... 197

		PAGE
Table of the safe Working Torsion for Cast-iron Shafts		200
Table of the safe Working Torsion for Wrought-iron Shafts		201

DIVISION IV.:—
Velocity of Shafts	204
Table of the Diameter of Wrought-iron Shafting necessary to transmit with safety various Amounts of Force	205

DIVISION V. ON JOURNALS:—
Length of Journals	207
Ultimate Pressure per Square Inch on Journal	208
Form of Journals	208

DIVISION VI. FRICTION:—
Laws of	209
Rennie's Table of Coefficients of Friction under Pressures increased continually up to Limits of Abrasion	212

DIVISION VII. LUBRICATION:—
Lubricants	213
Method of effecting complete Lubrication	215

CHAPTER IV.

ON COUPLINGS FOR SHAFTS AND ENGAGING AND DISENGAGING GEAR.

COUPLINGS:—
Primitive Cast-iron square Coupling-box	216
The Claw Coupling	217
Mr. Hewe's Coupling	218
The Disc Coupling	219
The Circular Half-lap Coupling	219

	PAGE
Rules for the Proportions of the Half-lap Coupling.	220
The Cylindrical Butt-end Coupling.	220

DIVISION VIII. DISENGAGING AND RE-ENGAGING GEAR :—

Throwing Wheels out of Gear with an Horizontal Lever	222
Throwing Wheels out of Gear with a Standard or Plummerb-lock and Movable Slide	223
Disengaging Machinery by the Fast and Loose Pulley	225
Disengaging Machinery with the Sack Teagle Motion	226
Callendering Marine Friction Clutch	227
Friction Cones	227
Friction Discs	228
Friction Couplings	229
Disengaging and Re-engaging Clutch	230
Two other Forms of ditto	231
Mr. Bodmer's Clutch	236

DIVISION IX. HANGERS, PLUMMER-BLOCKS, ETC., FOR CARRYING SHAFTING :—

Pedestal for supporting Shafting on the Floor	238
Pedestal for bolting Shafting to a Wall	239
Hanger for suspending Shafting from a Beam in the Ceiling	240
Hanger for suspending Shafting from the Floor	241
Hanger where great Strength is required	242
Hanger to connect two or three Ranges of Shafting.	244
Method of connecting Ranges of Shafting at Right angles to each other by means of Plummer-blocks.	246
Table of the Diameters, Pitch, Velocity, etc., of Spur Fly-wheels of the new Construction	250

MAIN SHAFTS :—

	PAGE
Material, Diameter, etc.	251
Description of the Main Vertical Shafts	252
Description of the Method of Gearing the Saltaire Mills	252
Method adopted to lessen the Friction on the Foot of the Vertical Shaft	256
Present Method by Bevel Wheels	258
Transmission of Power to Machinery at Obtuse Angles by the Universal Joint	259
Table of the Length, Diameter, etc., of Couplings, Coupling-Boxes, etc.	261

PRINCIPLES OF MECHANISM.

CHAPTER I.

GENERAL VIEWS.—LINK-WORK.—WRAPPING CONNECTORS.—
WHEEL-WORK.—SLIDING CONTACT.

I. GENERAL VIEWS RELATIVE TO MACHINES.

Definitions and Preliminary Expositions.

1. Mechanism may be defined as the combination of parts or pieces of a machine whereby motion is transmitted from the one to the other.

2. When a body, or any piece of mechanism, moves in a straight line it is said to have a *rectilinear* motion, and when it moves in a curved line it is said to have a *curvilinear* motion. When a point moves constantly in the same *path*, it is said to have a *continuous motion*, but if it moves backwards and forwards it is said to have a *reciprocating motion*. We may have *reciprocating rectilinear motions* as well as *reciprocating curvilinear motions*.

If a body moves over equal spaces in equal intervals of time, it has a *uniform motion;* but if it moves over unequal spaces in equal intervals of time, it has a *variable motion.*

3. The velocity of a body is the *rate* at which it moves. In uniform motion the velocity is constant; but in variable motion the velocity continually changes. If the velocity of a body increase it is said to be accelerated, and if the velocity decrease it is said to be retarded.

The motion of a body is said to be *periodical* when it undergoes the same changes in the same intervals of time.

4. In order to express the velocity of a body, we must have a certain number of units of space passed over in a certain unit of time. It is customary to take a foot as the unit of space, and a second as the unit of time.

In uniform motion, the space passed over is equal to the product of the velocity by the time. Thus, let s be the space in feet, t the time in seconds, and v the velocity per second; then

$$s = v\,t \ldots (1)$$

which expresses the general relation of space, time, and velocity, in uniform motions. Any two of these elements being given the remaining one may be found; thus we have

$$v = \frac{s}{t} \cdots (2), \text{ and } t = \frac{s}{v} \cdots (3).$$

5. If the velocity in one certain direction be taken as positive, then, that in the opposite or contrary direction will be negative.

6. If two wheels perform a revolution in the same time, their angular velocities are equal, whatever may be the dimensions of the wheels. The angular velocity of a revolving wheel or rod is the velocity of a point at a unit distance from the centre of motion. The wheel or rod will revolve uniformly when the angular velocity is uniform. If A be the angular velocity, r the radius of the wheel or length of the rod, v the velocity at this distance from the centre of motion; then

$$A = \frac{v}{r} = (1), \text{ and } v = A\,r \ldots (2).$$

7. The motion of wheels is conveniently expressed by the number of rotations which they perform in a given time. Thus, let n be the number of revolutions performed per min., the other notation being the same as in Art. 6; then

$$v = \frac{1}{30}\pi\,n\,r \ldots (1), \text{ and } n = \frac{30\,v}{\pi\,r} \ldots (2).$$

Or substituting A for $\frac{v}{r}$. See formula (1), Art. 6,

$$n = \frac{30\,A}{\pi} \ldots (3), \text{ and } A = \frac{1}{30}\pi\,n \ldots (4).$$

Hence the number of turns performed in a given time varies as the angular velocity.

The number of turns which two wheels respectively make in the same time is called their *synchronal* rotations. Let Q and q be the synchronal rotations of two wheels whose angular velocities are A and a, respectively; then $\dfrac{Q}{q} = \dfrac{A}{a}$; that is, synchronal rotations are in the ratio of the angular velocities.

Example.—Let a wheel whose radius is 6 ft. perform 50 revolutions per min., required 1st, the velocity of its circumference, and 2nd, its angular velocity.

Here, by eq. (1), $n = 50$, and $r = 6$, then
$$v = \frac{1}{30} \times 3{\cdot}1416 \times 50 \times 6 = 31{\cdot}416 \text{ ft. per sec.}$$

And, by eq. (4), $A = \dfrac{1}{30} \times 3{\cdot}1416 \times 50 = 5{\cdot}236$.

8. If v and v be the velocities of two parts of a piece of mechanism, then $\dfrac{V}{v}$ is the *velocity ratio* of these parts. Let s and s be the corresponding spaces described in the same time, then when the motion is uniform
$$\frac{V}{v} = \frac{S}{s} = \text{a constant,}$$
that is, when the velocities are uniform, the velocity ratio is constant.

9. If the velocity ratio of the two parts remains constant, then however variable the velocities themselves may be, we still shall have $\dfrac{V}{v} = \dfrac{S}{s}$, where s and s are the entire spaces described in the same interval of time.

GENERAL VIEWS. 17

10. When a body moves with a variable motion, its velocity at any instant is determined by the rate at which it is moving at that particular instant, that is, by the space which it would move over in one second, supposing the motion which it then has to remain constant for that time.

Variable motions may be graphically repre-

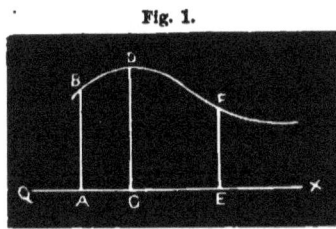
Fig. 1.

sented, by taking the abscissa of a curve equal to the units of time, and the ordinates equal to the units of the corresponding velocities. Thus let A B be equal to the units of velocity at the commencement of the motion; A C the units in interval of time, C D the units in the corresponding velocity; and so on; then the area of the curved space A B D F E will be equal to the space described in the interval of time represented by A E.

If the motion be uniform the curve B D F will become a straight line parallel to O X, and the space described in any given time will be represented by the area of a rectangle, whose length is equal to the units of time, and breadth equal to the units of velocity.

If the motion be uniformly accelerated or retarded, the curve B D F will become a straight line inclined to the axis O X, and the space described in this case, will be represented by the area of a

2*

trapezoid, whose base is equal to the units of time and parallel sides respectively to the velocity at the commencement and end of that time.

11. THE PARTS OF A MACHINE.—A machine conists of three important parts.

(1.) The parts which receive the work of the moving power—these may be called RECEIVERS of work.

(2.) The parts which perform the work to be done by the machine—these may be called WORKING PARTS, or more simply, OPERATORS.

(3.) The mechanism which transmits the work from the receivers to the working parts or operators—these pieces of mechanism may be called COMMUNICATORS OF WORK, or the TRANSMISSIVE MACHINERY.

The form of the mechanism must always be determined from the relation subsisting between the motions of the receivers and operators.

If there were no loss of work in transmission (from friction, etc.) the work applied to the receiver would always be equal to the work done by the operator. Thus, let P be the lbs. pressure applied to the receiver, and S the space in feet which it moves over in a certain time; p the lbs. pressure produced at the working part, and s the space in feet which it moves over in the sam time; then, neglecting the loss of work by friction we have—

Work applied to the receiver = work done upon the operator,

$$P \times S = P_1 \times S_1 \ldots (1).$$

However, it must be borne in mind, that the actual or useful work done by a machine is always a certain fractional part of the work applied; this fraction, determined for any particular machine, is called the modulus of that machine. If m be put for this modulus, then we have from eq. (1)

$$m \times P \times S = P_1 \times S_1 \ldots (2).$$

In treating of the motion of these parts of a machine it is generally most convenient to find an expression for their proportional velocities. Thus, let V be the velocity of the receiver, and V_1 that of the operator; then $\dfrac{V}{V_1}$ is their velocity ratio. See Art. 8.

It must be observed, that this *velocity ratio* is not at all effected by the *actual velocities* of the parts, provided the velocity ratio of the mechanism be constant for all positions. In the more ordinary pieces of mechanism (such as common toothed wheels, wheels moved by straps, levers, etc.) the velocity ratio is constant, that is to say, it remains the same for all positions of the mechanism.

In eq. (1) s may be taken as the velocity of the power P, estimated in the direction in which it acts, and s_1 that of the resistance P_1; then this equality becomes—

$$P \times V = P_1 \times V_1 \ldots (3),$$

or, $\dfrac{P_1}{P} = \dfrac{V}{V_1} =$ the velocity ratio $\ldots (4)$.

Now $\dfrac{P_1}{P}$ is called the advantage gained by the machine, or the number of times that the resistance moved is greater than the power applied. Hence the advantage gained by a machine, irrespective of friction, etc., is equal to the velocity of the power divided by the velocity of the resistance, or the velocity ratio of the power and resistance.

This is called the *principle of virtual velocities*. Workmen express this dynamic law by saying, " What is gained in power is lost in speed."

12. The DIRECTIONAL RELATION of the motion of the receiver and the operator admits of every possible variation. It may be constant or it may be variable. By the intervention of mechanism rectilinear motion may be converted into curvilinear motion, and conversely; reciprocating rectilinear or circular motion, may be converted into continuous circular motion, and conversely; and so on to the various possible combinations of which the cases admit. These directional changes are so important, in a practical point of view, that some eminent writers on mechanism have made

them the basis of the classification of mechanism. But, however eligible in a practical point of light such a classification may be, there is complexity in its application, which renders it less suitable for scientific purposes than that method of classification which is based upon the nature or mode of action of certain elementary pieces of mechanism which enter, more or less, into every mechanical combination.

Elementary Forms of Mechanism.

13. In analysing the parts of a machine we find motion transmitted by jointed rods or links, by straps and cords, by wheels rolling on other wheels, and by pieces of various forms sliding or slipping on other pieces. Hence we have the following elementary forms of mechanism:

(1.) Transmission of motion by jointed rods,— LINK-WORK.

(2.) By straps, cords, etc.,—WRAPPING CONNECTORS.

(3.) By wheels or curved surfaces, revolving on centres, rolling on each other,—WHEEL-WORK.

(4.) By pieces of various forms, sliding or slipping on each other,—SLIDING-PIECES.

14. The velocity ratio, as well as the directional relation, in an elementary piece of mechanism may be either constant or varying. The number of combinations of which these elementary pieces admit, is almost unlimited. The eccentric wheel

is a combination of sliding pieces and link-work. The common crane is a combination of wheel-work, link-work, and wrapping connectors; and so on to other cases.

A train of mechanism must be supported by some frame work; the train of pieces being such, that when the receiver is moved the other pieces are constrained to move in the manner determined by the mode of their connection. Revolving pieces, such as wheels and pulleys, are so connected with the frame that every portion of them is constrained to move in a circle round the axis; and sliding pieces are constrained to move in straight lines by guides.

Mechanism is to a great extent a geometrical inquiry. The motion of one piece in a train may differ, both in kind and direction, from the motion of the next piece in the series: these changes are effected by the geometrical construction of the pieces, as well as by their mode of connection. The investigation of the law of these changes constitutes one of the chief objects of the principles of mechanism.

II. ON LINK-WORK.

15. If a bent rod or lever A C B turn upon the centre C, the velocities of the extremities A and B

will be to each other in the ratio of their distances from the centre of motion c, that is,

$$\frac{\text{velocity A}}{\text{velocity B}} = \frac{\text{circum. cir. A Q}_1}{\text{circum. cir. B Q}} = \frac{\text{A C}}{\text{B C}}$$

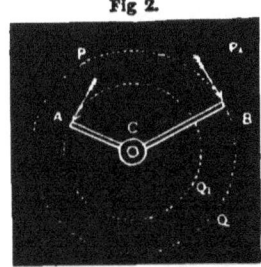

Fig 2.

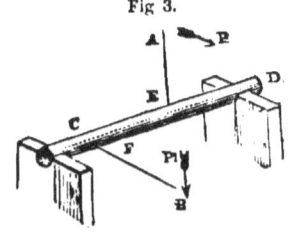

Fig 3.

It is not necessary that the arms A C and B C should be in the same plane. Thus let C D be an axis round which the arms A E and B F revolve, then,

$$\frac{\text{velocity A}}{\text{velocity B}} = \frac{\text{perpend. dist. A from the axis}}{\text{perpend. dist. B from the axis}}$$

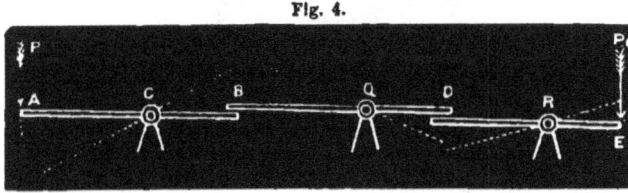

Fig. 4.

16. Let A B, B D, D E, be a series of levers turning on the fixed centres C, Q, and R; then when the arcs, through which the extremities A and E are moved, are small the velocity ratio will be expressed by the following equality:—

$$\frac{\text{velo. A}}{\text{velo. E}} = \frac{\text{A C.}}{\text{B C.}} \frac{\text{B Q.}}{\text{D Q.}} \frac{\text{D R}}{\text{E R}},$$

that is to say, *the velocity ratio of* P *and* P_1 *is found by taking the product of the lengths of the arms lying toward* P, *and dividing by the product of those lying toward* P_1.

17. To find *the velocity ratio of the rods* A B *and* C D, *turning on the fixed centres,* A *and* D; *and connected by the link* B C.

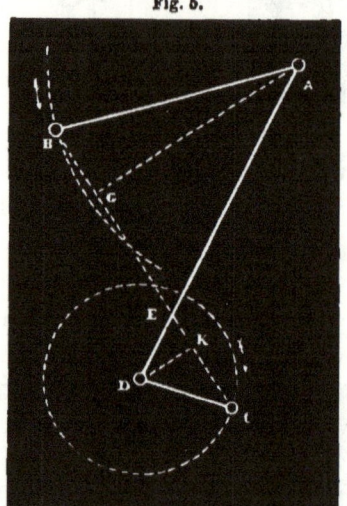

Fig. 5.

Through the centres A and D, draw the straight line D E A, cutting C B in E; and from A and D let fall the perpendiculars A G and D K upon C B, or it may be upon C B produced. Then

$$\frac{\text{ang. velo. D C}}{\text{ang. velo. A B}} = \frac{\text{A G}}{\text{D K}} \ldots (1);$$

that is to say, *the angular velocities of the rods* D C *and* A B *are to each other in the inverse ratio of the perpendiculars let fall from their respective axes upon the direction of the link.*

Similarly we also have:

$$\frac{\text{ang. velo. D C}}{\text{ang. velo. A B}} = \frac{\text{A E}}{\text{D E}} \ldots (2);$$

that is to say, *the angular velocities of the rods* D C *and* A B *are to each other in the inverse ratio of the segments into which the link divides the line joining their axes.*

These velocity ratios are obviously varying, depending upon the relative positions of the rods.

18. THE CRANK AND GREAT BEAM.—Let A B represent one half of the great beam of a steam-engine, D C the crank, and B C the connecting rod. Putting β for the angle D C B, and β for the angle A B C; then

$$\frac{\text{velo. crank}}{\text{velo. beam}} = \frac{\sin \beta_1}{\sin \beta} \ldots (1).$$

When the connecting rod B C is very long as compared with the length of the crank D C, then β is nearly constant, being nearly equal to 90°, in this case, eq. (1) becomes

$$\frac{\text{velo. crank}}{\text{velo. beam}} = \frac{1}{\sin \beta} \ldots (2).$$

The crank must be in the same straight line with the connecting rod, at the highest and lowest points of the stroke of the beam, and then $\beta = 0$. In these positions the crank is said to be at its dead points.

The velocity ratio, expressed by eq. (2), will be a maximum when $\beta = 0$, that is, the velocity of the crank will be a maximum when it is in its dead points. When $\beta = 90°$, or when the crank is at right angles to the connecting rod, then the velocity of the crank is a minimum.

If R = A B, or one-half the length of the great beam; r = D C, the length of the crank; and A = the angular oscillation of the beam, or the whole angle described by the beam in one stroke; then

$$r = R \sin \frac{A}{2} \ldots (3)$$

which expresses the length of the crank in terms of the radius of the beam and angle of its stroke.

A *double oscillation* of the beam produces *one complete rotation* of the crank, or conversely, taking the crank as the driver, *each rotation* of the crank produces *a double oscillation* in the beam.

From eq. (1) it follows, that the velocity of the crank is equal to the velocity of the beam, when $\beta = \beta_1$ or angle D C B is equal to angle A B C; that is, when the position of the crank is parallel to that of the beam.

By this form of the crank the *reciprocating circular* motion of the extremity of the beam is changed into a *continuous circular motion;* and conversely a *continuous circular motion* is changed into a *reciprocating circular motion*.

19. *To determine the various relations of position and velocity of the* CRANK *and* PISTON *in a locomotive engine.*

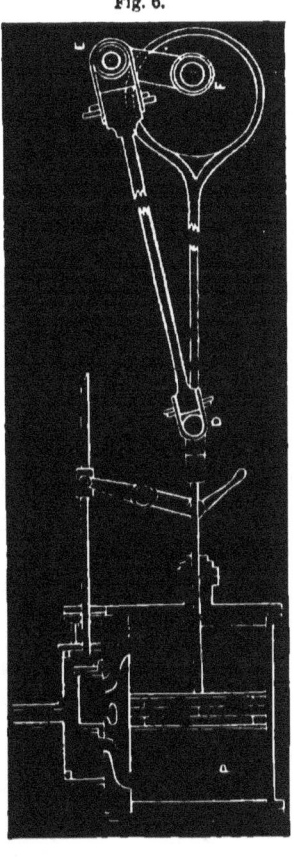

Fig. 6.

Here the connecting rod, D E, is attached to the extremity of the piston rod, P D, and *the length of the stroke of the piston is equal to double the length of the crank*, F E. Moreover, the centre, F, of the crank is in the same straight line with the axis of the cylinder or the direction of the piston rod.

Let $l =$ D E, the length of the connecting rod;

$l_1 =$ P D, the length of the piston rod;

$r =$ F E, the length of the crank;

$k =$ F D, the varying distance of the extremity of the piston rod from the axis of the crank;

$h =$ the corresponding height of the stroke of the piston;

$\theta =$ the varying angle, F E D, which the crank forms with the direction of the connecting rod.

(1.) *The velocity ratio of the crank and piston* is expressed by the following equality:

$$\frac{\text{velo. crank}}{\text{velo. piston}} = \frac{k}{l \sin \theta} \ \ldots (1), \text{ or}$$

$$= \frac{1}{\sin \beta} \ \ldots (2),$$

where β in eq. (2) is put for angle E F D; that is, the angle which the crank makes with the direction of the piston rod.

This latter form of the expression is the same as that given in eq. (2), Art. 18.

(2.) When the piston is at the bottom point of its stroke, its distance from F $=$ F E $+$ E D $+$ D P $= r + l + h$; also F D $=$ F E $+$ D E $= r + l$.

When the piston is at the middle point of its stroke, then F D $=$ E D; that is to say, in this position of the piston D E F will be an isosceles triangle.

(3.) *The position of the crank at any point of the stroke of the piston is determined by the two following general equations:—*

$$k = r + l - h \ \ldots (3).$$

$$\cos \theta = \frac{r^2 + l^2 - (r + l - h)^2}{2\,r\,l} \ \ldots (4).$$

When the piston is at the middle point of its stroke, then $h = r$, and eq. (4) becomes

$$\cos \theta = \frac{r}{2\,l} \ \ldots (5).$$

When the crank is at right angles to the connecting rod, $\theta = 90°$, and then we find from eq. (4),

$$h = r + l - \sqrt{r^2 + l} \ldots (6).$$

This expression is, obviously, less than r, or half the whole stroke of the piston. Hence it appears that the crank is at right angles with the connecting rod, before the piston has attained the middle point of its upward stroke.

20. Fig. 7 shows how a rotation of the axis A is transmitted to another C, by means of the two equal cranks A B and C D, connected by the connecting rod D B, whose length is equal to the distance A C, between the two axes. In all positions of the cranks, the figure A B C D will be a parallelogram, and the velocity of D will always be equal to the velocity of B, and the motion of the axis C will be exactly the same as that of the axis A.

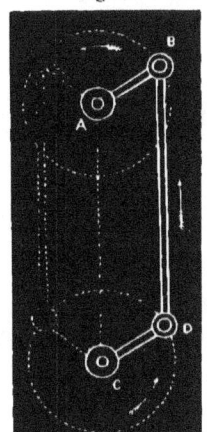

Fig. 7.

21. Two sets of cranks may be placed upon the axes, having the cranks on each axis at right angles to each other, similar to the mode of connecting the wheels of a locomotive engine, as shown in fig. 8, where the cranks are formed by bending, or loops made in the axes. These axes must be parallel to each

other, and the connecting rods must also be of equal lengths.

The advantage of this combination consists in maintaining a constant moving pressure, by which means an equable motion is sustained without the aid of the inertiæ of the machinery.

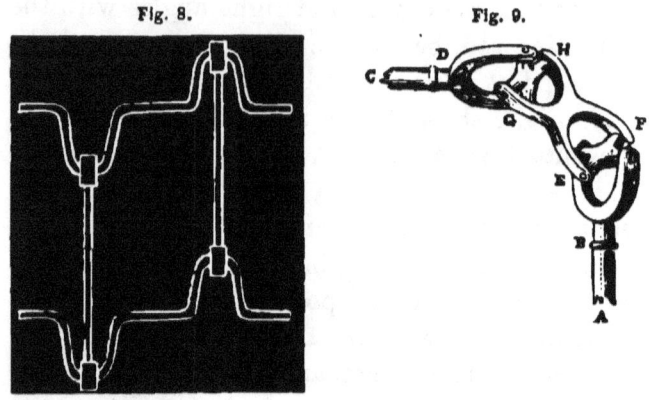

Fig. 8.   Fig. 9.

22. The double universal joint, represented in fig. 9, furnishes another example of link-work, for transmitting motion from one axis to another axis. This useful piece of mechanism should be constructed, so that the extreme axes, A B and C D, would meet in a point, if produced, and the angles which they respectively make with the central line of the intermediate piece, E F H G, shall be equal to each other.

LINK WORK. 81

Fig. 10.

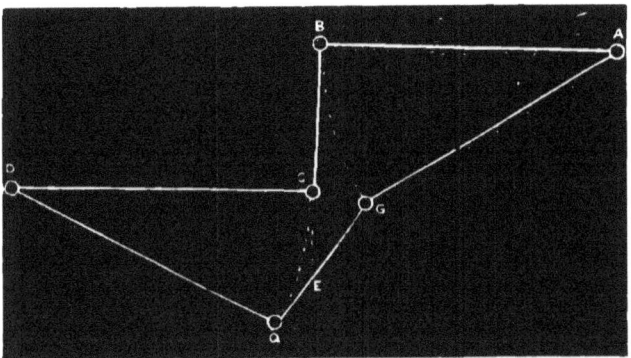

TO CONSTRUCT WATT'S PARALLEL MOTION.

23. This beautiful and useful piece of mechanism is formed by a combination of link-work.

Let A B and C D (see figs. 10 and 11) be two rods, turning on the fixed centres A and D, and connected together by the short link C B; then when motion is given to the rods, there is a certain point, E, in the link C B, which will move,

Fig. 11.

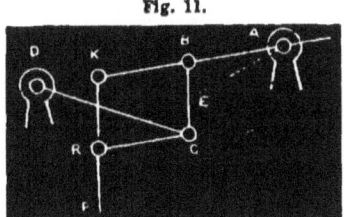

or very nearly move, in a straight line. In matter of fact the *path*, or *locus*, of this point is a curve of the fourth degree; but when the motion of the rods is limited, and their lengths are considerable, as compared with the length of their connecting link, this path becomes almost exactly a straight line.

In fig. 11, C B K R is a parallel frame of links; to the joint R is attached the piston rod R P of the steam engine; and to the point E is attached the piston rod of the air-pump.

(1.) *To find the point* E (see fig. 10) *to which the air-pump rod must be attached, having given the radius rod* C D, *the link* C B *or* Q G, *and the rod* A B *or* A G *forming a part of the great beam.*

Let D Q, A G be an extreme position of the rods. Let the rods be moved to the position A B C D, where the link C B is perpendicular to A B and D C. Produce B C, meeting the link Q G in the point E; then E will be that point of the link which will most nearly move in a vertical straight line. The ratio of Q E to G E is generally expressed by the following equality:—

$$\frac{QE}{GE} = \frac{R}{r} \times \left(\frac{r \sin \frac{a}{2}}{R \sin \frac{A}{2}} \right) \ldots (1);$$

where $R = AB$, $r = DC$, $a =$ angle C D Q, and $A =$ angle B A G.

Practically the link Q G or C B deviates very little from the vertical; and the angles a and A are small; hence, $r \sin \frac{a}{2} = R \sin \frac{A}{2}$ very nearly; in this case, therefore, eq. (1) simply becomes

$$\frac{QE}{GE} = \frac{R}{r} \ldots (2);$$

and from this equality we readily find,

$$GE = \frac{DQ \times GQ}{DQ + AG} \quad \ldots (3),$$

which gives the position of the point E, as required.

When $DQ = AG$, then $GE = \dfrac{GQ}{2}$, that is to say, in this case, the point E is at the middle of the link QG or CD.

Example.—Let AB or AG = 5 ft.; DC or DQ = 4 ft.; and CB or GQ = 1·5 ft.; then by eq. (3) we have,

$$GE = \frac{4 \times 1\cdot 5}{4 + 5} = \frac{2}{3} \text{ ft.}$$

(2.) *To find the length of the radius rod* DC (see fig. 13), *when the divisions* AB *and* BK, *on the beam are given.*

In this case,

$$\text{The radius rod, } DC = \frac{AB^2}{BK} \quad \ldots (4).$$

When $AB = BK$; then $DC = AB$; that is, in this case, the radius rod will be equal to the division AB on the beam.

Example.—Let AB = 6 ft., and BK = 4 ft.; then by eq. (4) we have,

$$\text{The radius rod, } DC = \frac{6^2}{4} = 9 \text{ ft.}$$

To multiply Oscillations by means of Link-work.

24. Fig. 12 represents a system of links B A C, C D, and D E, turning on the fixed centres A and E, and having the arms A B and A C united to the same centre A. The construction is such, that while the rod A B makes a *single* oscillation from B to I, the rod E D will make a *double* oscillation, viz., from D to F, and back from F to D. The oscillations of A B are produced by the rotation of a crank (see Art. 17), or by any other means.

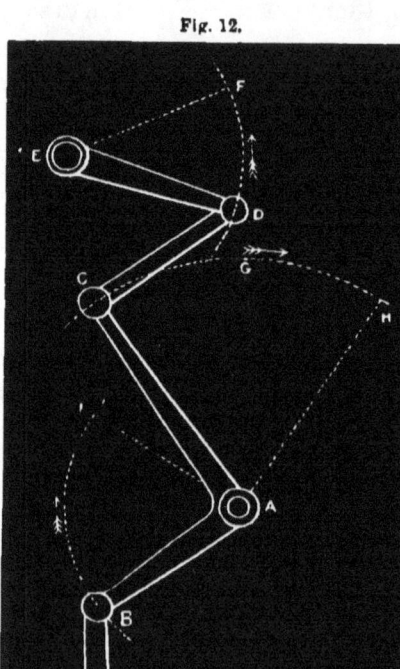

Fig. 12.

The conditions of the construction may be stated as follows:

Given the lengths of the arms A C and E D, the lengths or angles of their oscillations, and the length of the connecting link C D, to construct the mechanism, so that the rod

E D shall perform two oscillations whilst A B makes one.

Let B A C be the position of the bent lever at the commencement of the upward oscillation. Draw A I and A H, making the angles B A I and C A H each equal to the angle of the oscillation. From A as a centre, with A B and A C as radii, describe the arcs B I and C H. Through A draw A G F bisecting the angle C A H cutting the arc C H in G. On A G F take A F equal to the sum of the rods A C and C D, and make F D equal to the given length of the oscillation of E D. From D and F as centres, with a radius equal to the length of the rod E D, describe circles, cutting each other in E; then E will be the centre of the rod E D, which will perform two oscillations, whilst the rod A B makes one.

When A B and A C are in the middle points of their oscillations, the rod E D will have the position E F, that is, it will have performed a complete upward oscillation. When A B and A C have performed the remaining halves of their oscillations, the rod E F will have returned to the original position, that is, it will have performed a complete downward oscillation.

In like manner the oscillations may be further multiplied, by connecting E D with another series of links.

*To produce a Velocity which shall be rapidly
retarded, by means of Link-work.*

25. In fig. 13, R A C and E D represent two rods, turning on fixed centres A and E, and connected by a link C D; the rod E D is supposed to oscillate uniformly between the positions E D and E F. Now the construction is such as to produce a rapidly retarded motion of the rod R C in moving

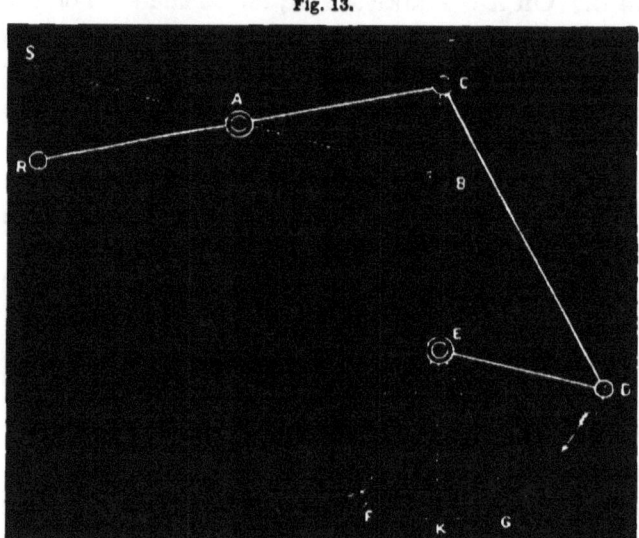

Fig. 13.

from the position R A C to the position S A B, and conversely.

The conditions of the construction may be stated as follows:

Given the rods E D and D C in position and magnitude, the angle of oscillation D E F, and the length of the rod A C, to construct the mechanism.

Bisect the arc D F in G, and then bisect the arc F G in K; through the points K and E, draw the straight line K E C; from D and K as centres, with a radius equal to the length of the link D C, describe arcs, cutting K E C in the points C and B; from B and C as centres, with a radius equal to the length of the rod A C, describe arcs cutting each other in the point A; then A will be the centre of the rod A C.

When the rod E D arives at the position E G, the rod R A C will have the position S A B *very nearly*, and it will have moved with a rapidly retarded motion. During the remaining half of the oscillation G F, the rod S A B will remain, virtually, stationary.

This piece of mechanism was first employed by Watt for opening the valves of the steam engine.

To produce a Reciprocating Intermittent Motion by means of Link-work.

26. A B and C D (fig. 14.) are two rods, turning on the fixed centres A and D, and connected by a link B C. The rod A B is made to oscillate between the positions A B and A I, by means of a crank and connecting rod. The construction of the mechanism is such, that the rod D C will oscillate between the positions D C and D F, but with an intermittent motion.

38   PRINCIPLES OF MECHANISM.

The conditions of the construction may be stated as follows:

Given the rods A B, B C, and C D in position and magnitude, to construct the mechanism.

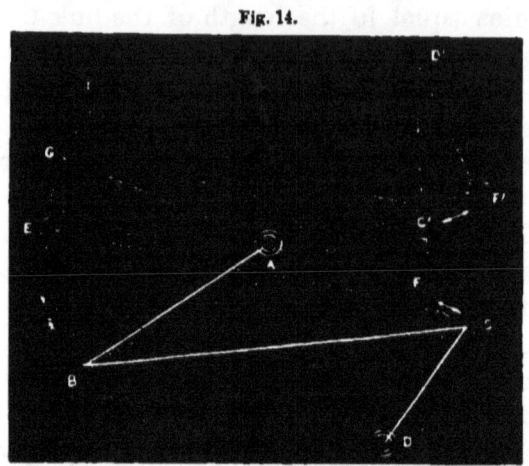
Fig. 14.

From A as a centre, with the radius A B, describe the arc B I; through C and A draw the straight line C A G, meeting the arc in G; make G E equal to one third the arc G B, and on the arc take G I equal to G E; on the line G A C take G F equal to B C; then half the chord B I will give the length of the crank, and C F will be the arc through which the rod D C oscillates.

Bisecting the angle B A E, &c., the position of the rod D' C' is found, which being connected with B, by the link B C', will oscillate exactly in a contrary manner to that of the rod D C, that is to say, when D C is stationary D' C' will be in motion, and conversely.

When the point B arrives at E, the rod D C will have completed, practically, its oscillation, and there it will remain stationary until the rod, turning on the centre A, returns from the position A I to A E.

The Ratchet-wheel and Detent.

27. In fig. 15, A represents the ratchet-wheel, and D the detent, falling into the angular teeth of the ratchet, thereby admitting the wheel to revolve in the direction of the arrow, but at the same time preventing it from revolving in the opposite direction.

Fig. 15.

In certain kinds of machinery, the action of the moving force undergoes periodic intermissions; in such cases the ratchet and detent are used to prevent the recoil of the wheels, and sometimes to give an intermittent motion to the wheel, as in the following example.

Intermittent Motion produced by Link-work connected with a Ratchet-wheel.

28. B E is a rod, turning on the fixed centre B, to which a reciprocating motion is given by the

connecting rod C of a crank, or by any other means; E F is a click, jointed to the rod B E at its extremity, and gives motion to the ratchet-wheel A. At each upward stroke of the rod B E, the click E F, acting upon the saw-like teeth of the ratchet-wheel, causes it to move round one or more teeth, and when the extremity F of the click is drawn back by the descent of the lever B E, it will slide over the bevelled sides of the teeth without giving any motion to the wheel, so that at every upward stroke of the rod C the ratchet-wheel will be moved round and it will remain at rest during every downward stroke of the rod. Thus the reciprocating motion of the connecting rod, C, will produce an intermittent circular motion in the axis A.

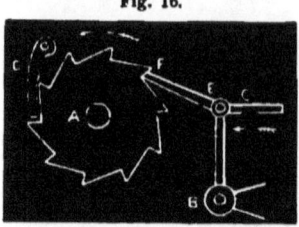

Fig. 16.

III. ON WRAPPING CONNECTORS.

29. When the moving force of the machinery is not very great, cords, belts, and other wrapping connectors, are most usually employed in transmitting motion from one revolving axis to another.

30. The *endless cord* or *belt* A B C D, represented in figs. 17 and 18, passes round the wheels, A B

WRAPPING CONNECTORS. 41

and C D, revolving on the parallel axes R K and Q F, and transmits motion from the axis Q F to the axis R K, with a constant velocity ratio. In all such cases the motion is entirely maintained by the frictional adhesion of the cord or belt to the surface of the wheel.

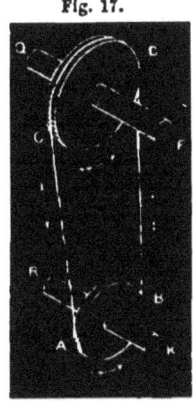
Fig. 17.

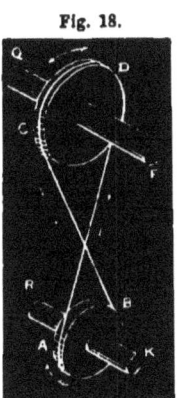
Fig. 18.

When the cord passing round the wheels is *direct*, as in fig. 17, the motions of the wheels take place in the *same direction*, and when the cords cross each other, as in fig. 18, the motions of the wheels take place in *opposite directions*.

If the wheel C D makes one revolution, then.

$$\text{No. revo. A B} = \frac{\text{circum. C D}}{\text{circum. A B}} = \frac{\text{radius C D}}{\text{radius A B}} \dots (1).$$

Or putting R and r for the radii of the wheels

C D and A B respectively, and Q and q for their respective synchronal rotations, then

$$\frac{q}{Q} = \frac{r}{R} \cdots (2).$$

Example.—If the radius of the wheel C D be 12 inches, and that of A B 9 inches, what will be the least number of entire revolutions which they must make in the same time?

Here, by eq. (2), we have

$$\frac{q}{Q} = \frac{R}{r} = \frac{12}{9} = \frac{4}{3}.$$

The fraction $\frac{12}{9}$ reduced to its least terms, is $\frac{4}{3}$, therefore the least number of synchronal rotations are 4 and 3, that is to say, whilst the wheel C D makes 3 rotations, the wheel A B will make 4.

31. Fig. 19 represents a system of three revolving axes, in which motion is transmitted from one to the other, by means of a series of belts.

Fig. 19.

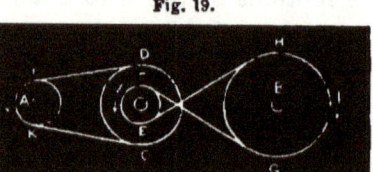

The belt being direct in the wheels A and D C, their axes will move in the same direction, but as the belt crosses in passing from D C to H G, their axes will move in opposite directions.

Here, whilst the axis B makes one rotation, the

$$\text{No. rotations A} = \frac{\text{rad. H G} \times \text{rad. D C}}{\text{rad. E F} \times \text{rad. I K}} \quad \ldots (1).$$

Or putting $R_1 =$ rad D C, $R_2 =$ rad. H G, &c., $r_1 =$ rad. I K, $r_2 =$ rad. E F, &c., and putting q and Q for the synchronal rotations of the first and last axes respectively; then

$$\frac{q}{Q} = \frac{R_1 \times R_2 \times R_3 \times \&c.}{r_1 \times r_2 \times r_3 \times \&c.} \quad \ldots (2).$$

Example.—In the mechanism represented in fig. 19, let $R_1 = 8$, $R_2 = 15$, $r_1 = 5$, $r_2 = 4$; required the least number of entire rotations performed in the same time by the axes A and B.

Here, by eq. (2) we have,

$$\frac{q}{Q} = \frac{8 \times 15}{5 \times 4} = \frac{6}{1}$$

that is, whilst the axis B makes one revolution, the axis A will make six.

32. In raising buckets from deep wells or from pits, a continuous cord coils round an axle or a drum wheel, as the case may be, the full bucket being attached to one end of the cord and the empty bucket to the other end; the rotation of the axle coils up the cord to which the full bucket is attached and at the same time uncoils the cord to which the empty one is attached, so that whilst the former is ascending the latter is descending.

Speed Pulleys.

33. Fig. 20 represents an arrangement of speed pulleys; A B and C D are two parallel axes upon each of which is fixed a series of pulleys, or wheels, adapted for a belt of given length, so that it may be shifted from one pair of wheels to any other pair, say for example, from the pair $a\ a_1$ to the pair $c\ c_1$. In order to suit this arrangement, if the belt be crossed, *the sum of the diameters of any pair of pulleys must be a constant quantity*, that is to say, it must be equal to the sum of the diameters of any other pair. By this contrivance, a change in the velocity ratio of the two axes is produced by simply shifting the belt from one pair to another.

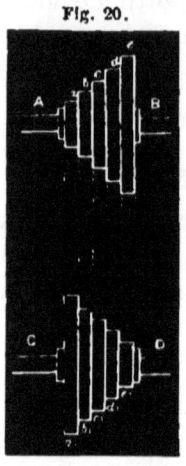

Fig. 20.

In practice it is customary to make the two groups of pulleys exactly alike, the smallest pulley of one being placed opposite to the largest of the other.

In a group of speed pulleys, let $s =$ the constant sum of the diameters of the driver and follower, $D =$ the diameter of the follower, and $Q\ q$ the number of their synchronal rotations respectively.

then $\dfrac{Q}{q} = \dfrac{d}{D}$, and

$$D = \frac{q \times s}{Q+q} \cdots (1);$$

$$d = \frac{Q \times s}{Q+q}, \text{ or more simply,}$$

$$= s - D \cdots (2).$$

Example.—Required the diameters of a pair of speed pulleys, when the sum of the diameters is 30 inches, and the driver makes two revolutions, whilst the follower makes three.

Here $s = 30$, $Q = 2$, and $q = 3$; then by eq. (1) and (2) we have

$$D = \frac{3 \times 30}{5} = 18 \text{ in.; and } d = 30 - 18 = 12 \text{ in.}$$

If the constant sum of the diameters of a group of 5 pairs of speed pulleys be 12 inches, and the diameters of the pulleys a_1, b_1, c_1, d_1, e_1, be 10, 8, 6, 4, and 2 inches respectively, then the diameters of the pulleys a, b, c, d, e, will be 2, 4, 6, 8, and 10 inches respectively; and as the strap is shifted from one pair of wheels to another, the relative velocities of the axes CD and AB will be as the numbers $\frac{1}{5}$, $\frac{1}{2}$, 1, 2, and 5.

34. It is customary to construct the pairs of speed pulleys so that the rotations of the follower may be increased or decreased in a certain geometric ratio. Thus, if r be this ratio, then for 5 pairs of speed pulleys we shall have the series of terms $\frac{1}{r^2}$, $\frac{1}{r}$, 1, r, r^2, for the different values of $\frac{Q}{q}$,

the ratio of the synchronal rotations of each pair. Or, generally, if n be the number of pairs, then $\dfrac{1}{r^{\frac{n-1}{2}}}, \dfrac{1}{r^{\frac{n-3}{2}}}, \ldots, r^{\frac{n-3}{2}}, r^{\frac{n-1}{2}}$, will be the different values of $\dfrac{Q}{q}$. In this case, let $D_1, D_2, \ldots, D_n =$ the diameters of the 1st, 2d, ..., and nth pulleys, respectively, on the driving axis; and these symbols, taken in a reverse order, will be the corresponding diameters of the pulleys on the driven axis; then $D_1 = \dfrac{S}{1+r^{\frac{n-1}{2}}}$, $D^2 = \dfrac{S}{1+r^{\frac{n-3}{2}}}$, and so on: moreover we have $D = S - D_1$, $D_{n-1} = S - D_2$, and so on.

Example.—To find the diameters of a set of 5 pairs of speed pulleys, so that values of $\dfrac{Q}{q}$ (the ratio of the synchronal rotations of the different pairs) shall have the common ratio of $\tfrac{2}{3}$, the constant sum of the diameters of each pair being 26 inches.

Here $r = \tfrac{2}{3}$, $n = 5$, and $S = 26$, then from the foregoing formulæ we find,

$$D_1 = \dfrac{26}{1+(\tfrac{2}{3})^2} = 18; \quad D_2 = \dfrac{26}{1+\tfrac{2}{3}} = 15\tfrac{3}{5};$$

$$D_3 = \dfrac{26}{1+(\tfrac{2}{3})^0} = 13; \text{ and so on.}$$

But the remaining diameters will be better found as follows:

$$D_6 = 26 - 18 = 8; \quad D_4 = 26 - 15\tfrac{3}{5} = 10\tfrac{2}{5}.$$

35. Two plain cones, having their axes parallel, as shown in fig. 21, will obviously answer the same purpose as the ordinary form of speed pulleys. The slant faces of the cones may be formed by any continuous curve; but with this condition—that the sum of the diameters at every position of the band shall be a constant.

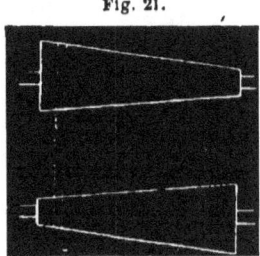

Fig. 21.

Guide Pulleys.

36. By the intervention of *guide pulleys* the direction of cords may be changed into any other direction. Thus, by means of the guide pulleys B and C, the motion of the cord in the direction C D is changed into the direction A B.

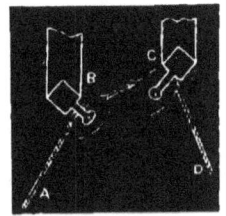

Fig. 22.

The cords D C and C B should be in the plane of the pulley C; and the cords C B and B A should be in the plane of the pulley B.

37. Two guide pulleys, E and H, may be employed to transmit motion from the wheel A to

the wheel B, when the axes of these wheels have any given direction.

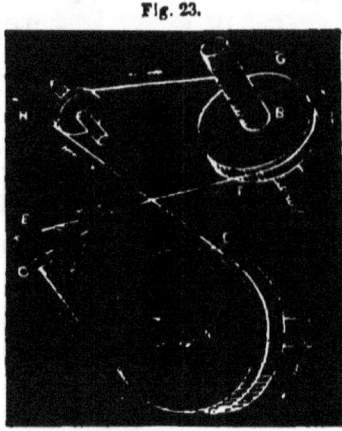

Fig. 23.

Let E H be the line where the planes, passing through the two wheels, intersect each other. In this line assume any two convenient points E and H; in the plane of the wheel A draw the tangents E C and H D; and in the plane of the wheel B draw the tangents E F and H G; then C E F G H D will be the path of the endless cord, which will be kept in this path by a guide pulley at E, in the plane of C E F, and another guide pulley at H, in the plane of D H G.

The relative velocities of the axes A and B depend entirely upon the ratio of the radii, A D and B G, of the two wheels. See Art. 30.

To prevent Wrapping Connectors from Slipping.

38. The slip of the band on the wheel, when it is not excessive, is in many cases rather an advantage than otherwise; but when motion is to be transmitted from one wheel to another according

to some given exact ratio, *gearing chains* of various forms are employed as the wrapping connectors.

39. In some cases the links of the gearing chain lay hold of pins or teeth formed upon the wheel, as shown in fig. 25. In other cases, the links of the gearing are joined together, something like a

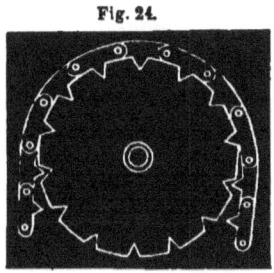
Fig. 24.

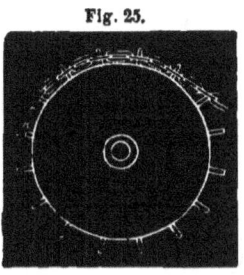
Fig. 25.

watch chain, and carry teeth which pass into certain notches made at corresponding distances on the edge of the wheel, as shown in fig. 24.

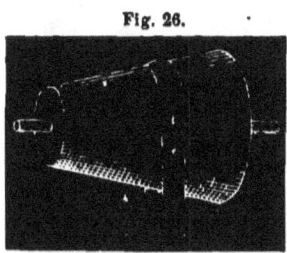
Fig. 26.

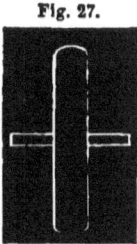
Fig. 27.

40. **When a belt moves a conical wheel, it always** happens that the belt gradually moves toward the broad end of the wheel; this is owing to the belt

being more stretched on that side than it is on the other.

41. This property enables us to construct a wheel so that a belt shall not shift on its edge; this is simply effected by making the edge to swell a little in the middle, as shown in fig. 27.

42. When two rollers have to make only a limited number of revolutions in each direction,

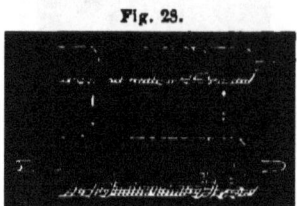

Fig. 28.

the slip of the cord may be prevented by having a cord coiled round each end of the rollers in opposite directions, so that while one cord is coiled on one extremity of the roller, the other cord is uncoiled from the other extremity, as shown in fig. 28.

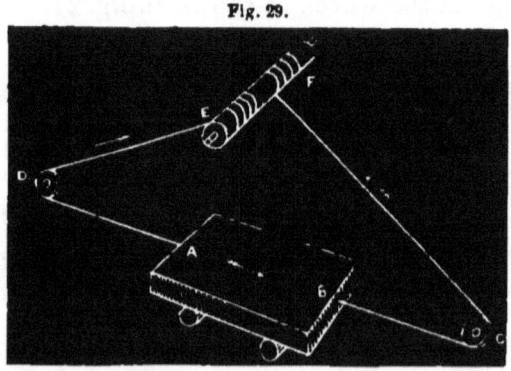

Fig. 29.

43. By a similar arrangement of cords on the cylinder E F (see fig. 29), a reciprocating motion

of this cylinder will produce a back and forward motion of the carriage A B.

Systems of Pulleys.

44. A system of pulleys must at least contain one movable pulley. When a wheel, forming a part of a system of wheels connected together by cords, has a progressive motion, it materially affects the velocity ratio of the receiver and the operator of the mechanism. There are a great many different systems of pulleys, but they all depend upon the different combinations of movable and fixed pulleys, and the different modes of reduplication of a cord.

45. In this system of pulleys there is one movable block and a single continuous cord with three duplications, so that whilst the moving force P acts by one cord, the movable block with its load is suspended by six cords; if W ascend one foot, each of these cords will be shortened one foot, and therefore the cord P will be lengthened six feet; that is to say, the velocity of P will be six times that of W.

46. In the system of pulleys represented in fig. 31, there are two distinct cords and two movable pulleys, A and B, making two duplications of cord;

then if A ascend one foot, B must ascend two feet, and the cord at P must be lengthened four feet; that is, the velocity of P will be four times the velocity of W.

Generally, if there are n moveable pulleys in such a system, then,

$$\text{velo. P} = 2^n \times \text{velo. W}.$$

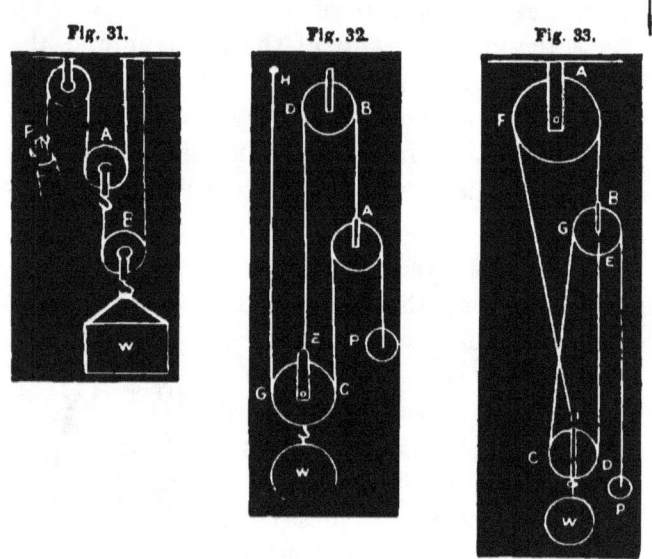

Fig. 31. Fig. 32. Fig. 33.

47. The system of pulleys represented in fig. 32, contains two movable pulleys, one fixed pulley, and two single cords. In this case the velocity ratio of P to W is as four to one.

48. Fig. 33 represents a similar system of pul

leys, in which the velocity ratio of P to W is as five to one.

In all these systems of pulleys the velocity ratios are constant.

49. In the compound wheel and axle, represented in fig. 34, the axle is made of different thicknesses, as at A and B, and a continuous cord coils round these parts in different directions, and passes round the wheel of the movable pulley D. In one revolution of the wheel C P,

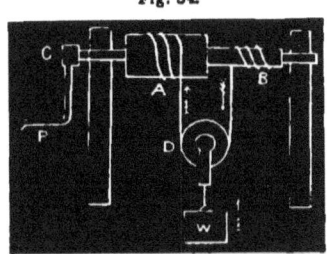
Fig. 34.

the space moved over by the pulley D is equal to half the difference of the circumferences of the axles A and B. Putting R_1 for the radius of the wheel C P, R for the radius of the axle A, and r for the radius of the axle B; then we have for the velocity ratio

$$\frac{\text{velo. P}}{\text{velo. W}} = \frac{2 R_1}{R - r}.$$

If $R_1 = 10$, $R = 4$, $r = 3\frac{3}{4}$; then

$$\frac{\text{velo. P}}{\text{velo. W}} = \frac{2 \times 10}{4 - 3\frac{3}{4}} = 80.$$

This piece of mechanism belongs to a class which produces what has been called *differential motions*, their object being to produce a slow and definite motion in a body by the most simple and practicable means.

TO PRODUCE A VARYING VELOCITY RATIO BY MEANS OF WRAPPING CONNECTORS.

50. To find the ratio of the angular velocities of two eccentric wheels, moved by a cord wrapping over each.

Let D C be a cord wrapping round the wheels,

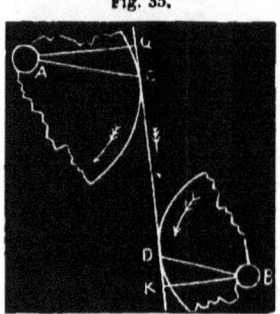

Fig. 35.

whose axes of motion are A and B; their line C D will be a tangent to the two curves forming the edges of the wheels. On D C produced let fall the perpendiculars A Q and B K; then the velocity of the cord, in this position of the wheels, will be equal to the velocity of the point Q, and at the same time it will also be equal to the velocity of the point K: hence we find,

$$\frac{\text{angular velocity A C}}{\text{angular velocity B D}} = \frac{\text{B K}}{\text{A Q}}, \ldots (1);$$

that is to say, the *angular velocities are inversely as the perpendiculars let fall upon the cord from the axes of motion.*

WRAPPING CONNECTORS.

51. Let B be a movable pulley suspended from the continuous cord P A B C, passing over a fixed pulley A, and attached to a point C in the same horizontal line with A. Let fall B D perpendicular to A C; then B C will always be equal to

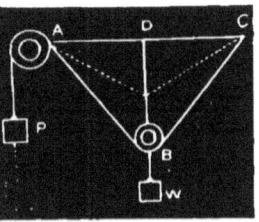
Fig. 36.

B A, and B will move in a vertical line B D. Hence we find,

$$\frac{\text{velocity P}}{\text{velocity W}} = 2 \times \frac{\text{B D}}{\text{B A}} \ldots (1).$$

This expression may be put in the following trigonometrical form:

$$\frac{\text{velocity P}}{\text{velocity W}} = 2 \times \cos \text{P A B} \ldots (2).$$

52. Fig. 37 represents a simple and ingenious contrivance for communicating a varying velocity to the axis B, by means of an endless band Q K C, passing over an eccentric wheel A, a pulley B K, and a *stretching pulley* C. The curve of the eccentric wheel A, must be such as to produce the varying velocity required. The

Fig. 37.

weight W, attached to the stretching pulley C, keeps

the band constantly stretched, so that whatever may be the velocity of the cord upon leaving the eccentric wheel, it communicates the same velocity to the circumference of the pulley B K. From the axis A let fall A Q perpendicular to the cord Q K; then by eq. (1), Art. 50, the velocity ratio may be expressed as follows:

$$\frac{\text{ang. velo. axis A}}{\text{ang. velo. axis B}} = \frac{\text{B K}}{\text{A Q}}.$$

Let the axis A revolve uniformly, and let the radius B K of the pulley be given; then

The ang. velo. axis B *will vary as the perpend.* A Q.

IV. ON WHEEL-WORK PRODUCING MOTION BY ROLLING CONTACT WHEN THE AXES OF MOTION ARE PARALLEL.

53. Two wheels, E and F, in contact with each other, revolve on the parallel axes A B and C D; now if the wheels are in contact in any one position, they will also be in contact in every other position, and their circumferences will roll upon each other, so that if the driver F revolve on its axis C D, it will communi-

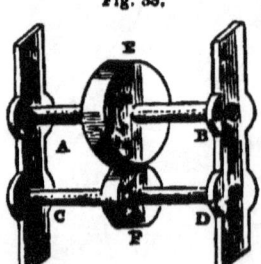

Fig. 33.

cate a rotatory motion to the follower E in a contrary direction, by the frictional adhesion of the parts successively brought in contact. The edges of these wheels must have the same velocity, and therefore their angular velocities will be inversely as their radii.

54. In order to render the transfer of motion perfectly exact, the edges of the wheels are formed into teeth, placed at equal distances from each other, so that when one wheel is turned, its teeth successively enter into the spaces formed on the edge of the other wheel. Thus, even with slight errors of construction, one wheel can not escape from the other, which may happen in the case of simple rollers.

Fig. 39.

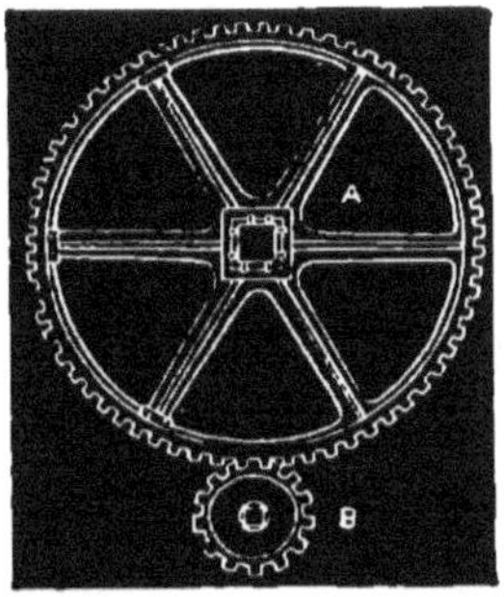

The numbers of teeth in the wheels, acting upon each in this manner, are in proportion to their radii. Thus, let the radius of the wheel A be 15 inches, that of B 6 inches, and let B contain 8 teeth; then

$$\text{No. teeth in A} = 8 \times \frac{15}{6} = 20.$$

Or, generally, if R and r be put for the radii of the

wheels, and N and n the number of their teeth respectively; then

$$\frac{N}{n} = \frac{R}{r} \ldots (1).$$

Hence angular velocities, as well as the synchronal rotations, of wheels, may be expressed in terms of their numbers of teeth; thus we have—

$$\frac{\text{ang. velo. A}}{\text{ang. velo. B}} = \frac{n}{N} \ldots (2);$$

also, $\dfrac{\text{synchronal rotation A}}{\text{synchronal rotation B}}$, or $\dfrac{Q}{q} = \dfrac{n}{N} \ldots (3).$

Example.—Required the least number of teeth in the wheels A and B, so that B shall make 105 revolutions per min. and A only 40.

Here by eq. (3), $\dfrac{n}{N} = \dfrac{40}{105} = \dfrac{8}{21}$;

that is, B will contain 8 teeth and A 21 teeth.

The form which must be given to the teeth of wheels, so as to maintain a perfect rolling contact, will be explained in another part of this work.

55. If the wheel A be the *driver* then B will be called the *follower*. Wheels acting in this manner are sometimes called *spur-wheels*. Small toothed wheels are called *pinions*; thus B may be called a pinion in relation to A.

56. Toothed wheels are said to be *in gear* when

their teeth are engaged together, and they are said to be *out of gear* when they are separated.

57. In the train of wheels represented in fig. 40,

Fig. 40.

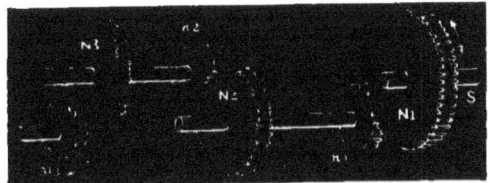

let N_1, N_2, N_3, &c., be the number of teeth in the *driving* wheels, and n_1, n_2, n_3, &c., the number in the *driven* wheels; $Q_1 =$ the no. of rotations of the first axis, $Q =$ the no. of the second axis, and so on, performed in the same time; then

$$\frac{Q_{m+1}}{Q_1} = \frac{N_1 . N_2 . N_3 . \ldots N_m}{n_1 . n_2 . n_3 . \ldots n_m} \ldots (1).$$

This equality may be expressed in language as follows:—*The ratio of the synchronal rotations of the last and first axes, is equal to the continued product of the number of teeth in the driving wheels divided by the continued product of the number of teeth in the driven wheels.*

Similarly we have,

$$\frac{Q_{m+1}}{Q_1} = \frac{Q_2}{Q_1} \times \frac{Q_3}{Q_2} \times \ldots \times \frac{Q_{m+1}}{Q_m} \ldots (2),$$

which may be expressed in language as follows:— *The ratio of the synchronal rotation of the first and last axes, is equal to the product of the separate synchronal ratios of the successive pairs of axes.*

The number of axes in this combination is always one more than the number of pairs of wheels.

It is evident, from eq. (1), that the drivers and followers may be placed in any order in a train of wheel-work without changing the velocity ratios of the first and last axes.

Example.—Let the number of pairs of drivers and followers be 3, that is, let $m = 3$, $N_1 = 16$, $N_2 = 15$, $N_3 = 14$, $n_1 = 7$, $n_2 = 6$, $n_3 = 5$; required the least number of synchronal rotations of the first and last axes in the train of wheels.

Here by eq. (1) we have—

$$\frac{Q_4}{Q_1} = \frac{16 \times 15 \times 14}{7 \times 6 \times 5} = \frac{16}{1};$$

that is, whilst the first axis makes one revolution, the last will make sixteen.

58. If the number of teeth in a driving wheel be some exact multiple of the number of teeth in the follower, then the same teeth will come into contact in every revolution of the driver. Thus if the driver contains 30 teeth and the follower 6, then the same teeth will come into contact at every revolution of the driver. This arrangement of teeth is preferred by the clock and watchmaker; but the millwright would add one tooth, called the HUNTING COG, to the large wheel, that is, he

would have 31 teeth in the driver and 6 in the follower, because 31 and 6, being prime to each other, and at the same time nearly in the same ratio as 30 and 6, the same pair of teeth would not come again into contact until the large wheel had made 6 revolutions, and the small one 31.

59. Eq. (3), Art. 53, enables us readily to find the number of revolutions which the wheels must make in order that the same teeth may come again into contact with each other; for it is only necessary to reduce the fraction $\frac{n}{N}$ to its least terms, and the denominator of this reduced fraction will give the number of revolutions of the driving wheel as required. Thus, let $N = 144$, and $n = 54$, then $\frac{Q}{q} = \frac{54}{14} = \frac{3}{8}$; that is, the driver must make 3 complete revolutions, or the follower 8, before the same teeth can again come into contact.

60. In a combination of wheels, whose motions are expressed by the equality $\frac{Q_2}{Q_1} = \frac{N_1 \cdot N_2}{n_1 \cdot n_2}$, an indefinite number of values may be assigned to the numbers of teeth, which shall produce a given synchronal ratio of the first and last axes; but if n_1 and n_2 be given, and N_1 and N_2 be comprised within certain given limits; then a limited number of values may be found for N_1 and N_2.

Thus, for example, let $\frac{Q_2}{Q_1} = 60$, $n_1 = n_2 = 8$, and the values of N_1 and N_2 not to exceed 100 nor to be less than 40.

Here we have—

$$\frac{N_1 \cdot N_2}{8 \times 8} = 60;$$

$$\therefore N_1 \cdot N_2 = 60 \times 64;$$

hence, N_1 may be 60 and N_2 may be 64; but in order to determine all the combinations, we must put the product, 60×64, into prime factors, and then distribute these factors into different groups answering to the limiting values of N_1 and N_2.

Here, $60 \times 64 = 2^8 \times 3 \times 5$; hence we have
1st combination, $(2^4 \times 3) \times (2^4 \times 5) = 48 \times 80$;
2d combination, $(2^5 \times 3) \times (2^3 \times 5) = 96 \times 40$;
3d combination, $2^6 \times (2^2 \times 3 \times 5) = 64 \times 60$.

61. When all the drivers contain the same number of teeth, and also the followers, then eq. (1), Art. 57, becomes

$$\frac{Q_{m+1}}{Q_1} = \left(\frac{N_1}{n_1}\right)^m \ldots (1).$$

By means of this formula we may readily determine the least number of axes requisite for producing a given synchronal ratio of rotation between the first and last axes, when the number of teeth in the drivers cannot exceed N_1 and the number in the followers cannot be less than n_1.

Find m, in eq. (1), equal to the highest whole number, which does not make the right member greater than the left; then the least number of

WHEEL WORK. 63

axes will be $m + 2$. But if m, a whole number, can be found so as to make the right-hand member exactly equal to the left, then in this case, the least number of axes will be $m + 1$.

Example.—Required the least number of axes in a train of wheels which shall cause the last axis to revolve 180 times as fast as the first axis, allowing that none of the drivers can contain more than 54 teeth, and none of the followers less than 9.

Here, we must find the greatest whole number for m, so that $\left(\dfrac{54}{9}\right)^m$ or $(6)^m$ shall not exceed 180. This value of m is obviously 2; and the least number of axes will be 4.

Idle Wheels.

62. The wheel c placed between two other wheels, A and B, does not affect the velocity ratio of these wheels; and hence the wheel c is called an idle wheel. This intermediate wheel, however, causes the wheels A and B to revolve in the *same direction*, whereas if A and B were in contact they would revolve in *opposite directions*.

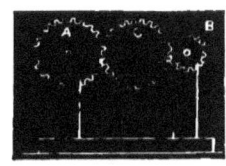

Fig. 41.

Annular Wheels.

63. Fig. 42 represents an annular wheel A, having its teeth cut on the internal edge of the annulus or rim. The toothed wheel B, revolving within the annular wheel A, causes it to revolve in the *same direction;* whereas two ordinary spur wheels revolve in *opposite directions.*

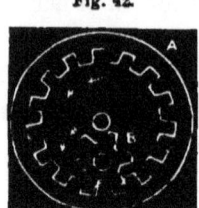
Fig. 42.

Concentric Wheels.

64. When two separate wheels revolve about the same centre of motion, they are called concentric wheels. The pinion D is fixed to the axis F E, whilst the concentric wheel C is fixed to a tube or cannon, N, which revolves freely upon the axis F E. The driving wheels, A and B, fixed to the parallel axis H G, communicate the relative velocities to the axis F E and to the cannon N.

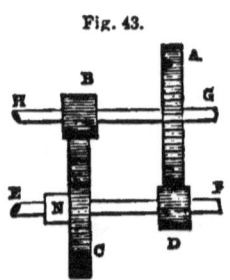
Fig. 43.

Wheel work when the axes are not parallel to each other.

65. When the axes of two wheels are not parallel to each other, motion is generally communicated from the one to the other by *bevel wheels* or

WHEEL WORK.   65

bevel gear. When the axes are perpendicular to each other, the *face wheel and lantern*, and the *crown wheel* are frequently employed.

Face Wheel and Lantern.

66. In fig. 44, F represents a *face wheel*, with its lantern L. Here motion is transmitted from the vertical axis A B to the horizontal axis A C. The teeth F on the face of the face wheel are called *cogs*, which are usually

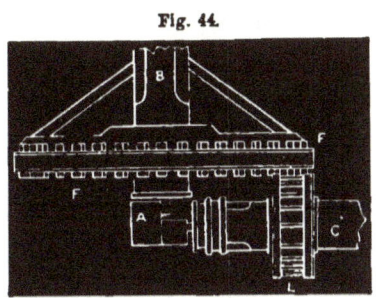

Fig. 44.

made of iron, whilst the round *staves* forming the teeth of the lantern, L, are made of hard wood. The axes A B and C D should, when produced, intersect in a point.

Crown Wheels.

67. Fig. 45 represents a *crown wheel* B, with its pinion A, having their axes at right angles to each other. The teeth of the crown wheel are cut on the edge of a hoop, the plane of which is at

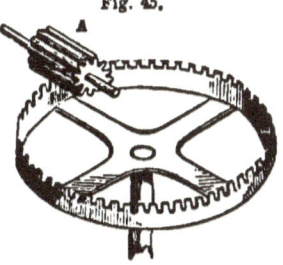

Fig. 45.

6*

right angles to its axis, and the pinion is thicker than wheels are commonly made.

CASE I. *To construct Bevel Wheels or Bevel Gear when the axes are in the same plane.*

68. Let A C and A B be two axes of rotation, in the same plane, and cutting each other in the point A. On these axes two right cones, A D F and A D E, may be formed, touching each other in the line A H D; and also two right frusta, D F G H and D H K E, of these cones may be formed.

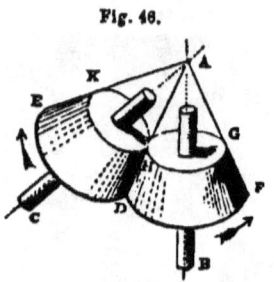

Fig. 46.

Now, if the frustum D F G H revolve on its axis B A, it will communicate, by rolling contact, a rotatory motion to the frustum D H K E upon its axis C A.

These frusta of cones will obviously perform their rotations in the same time as the ordinary spur wheels previously described.

On the surfaces of these frusta a series of equidistant teeth are cut, directed to the apex A of the cones, so that a straight line passing through the apex to the outline of the teeth upon the bases D F and D E of the frusta shall touch the teeth in every part as shown in the diagram.

Wheels cut in this manner are called *bevel gear.*

Two wheels of this construction will always

transfer motion, with a constant velocity ratio, from one axis to the other, provided these axes meet each other in a point, which point being always made the apex of the frusta forming the bevel of the wheels.

69. *General problem.*—Given the radii of two bevel wheels, and the position of their axes, to construct the frusta forming the wheels, the two axes being in the same plane.

Let A B and A C be the position of the axes cutting each other in A. Draw I J parallel to A B at a distance equal to the radius of the wheel on the axis A B; and draw M L parallel to A C, at a distance equal to the radius of the wheel on the axis A C, cutting the line I J

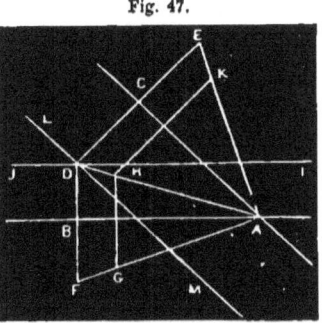

Fig. 47.

in the point D. From the point D, draw D B F perpendicular to A B, and D C E perpendicular to A C. Take B F equal to B D, and C E equal to C D. Join A E, A D, and A F. At a distance equal to the thickness of the wheel, draw H G parallel to D F, cutting A D in H; and through H, draw H K parallel to D E. Then D F G H and D H K E will be the frusta required.

CASE II. *To construct Bevel Gear when the axes are not in the same plane.*

70. This is usually done by introducing an intermediate wheel with two frusta formed upon it, one frustum rolling in contact with the driving wheel, and the other frustum in contact with the driven wheel.

71. Let A B and C D be the direction of the given axes; take A D as a third axis, meeting the axes A B and C D at any convenient points, A and D; then A will be the vertex of two rolling frusta of cones G and H, and D will be the vertex of two other rolling frusta of cones I and K.

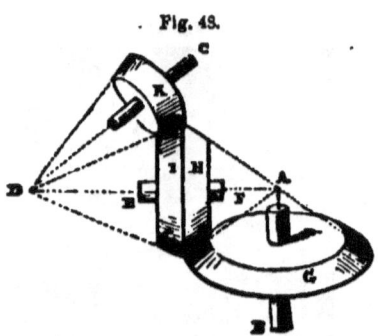

Fig. 43.

Whilst the intermediate axis, with its two frusta of cones, revolves, the teeth of the frustum H will have a rolling contact with the teeth of the frustum G, and at the same time the teeth of frustum I will have a rolling contact with the teeth of the frustum K; and thus motion will be transmitted from the axis A B to the axis C D with a constant velocity ratio.

Let Q_1 and Q_2 be the number of rotations performed by the axes A B and C D respectively in the same time; $N_1 =$ the number of teeth in the bevel wheel G; $n_1 =$ the number in the edge H; $N_2 =$ the number in the edge I; and $n_2 =$ the number in the bevel wheel K; then,

$$\frac{Q_2}{Q_1} = \frac{N_1 \cdot N_2}{n_1 \cdot n_2} \ldots (1),$$

which is similar to the expression given in eq. (1), Art. 57. When $n_1 = N_2$, then this equality becomes,

$$\frac{Q_2}{Q_1} = \frac{N_1}{n_2} \ldots (2).$$

In this case the intermediate bevel wheel, I H, may be regarded as an idle wheel.

VARIABLE MOTIONS PRODUCED BY WHEEL WORK HAVING ROLLING CONTACT.

72. Two curved wheels, E P and F P, having rolling contact, revolve on the axes A and B. In order that these wheels may roll on each other without slipping, or without producing any strain upon the axes A and B, these axes must always be in the line of contact A P B, and if the curve P E on the one wheel be equal to the curve P F on the other wheel, the sum of the lines A E and B F must always be equal to A B, the distance be-

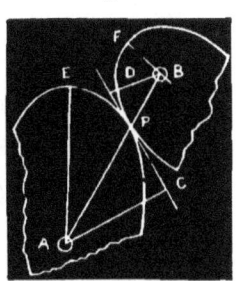

Fig. 49.

tween the centres of motion. Various curves may be constructed, having this property. For example, two equal ellipses, E P and F P, revolving on their foci, A and B, and having A E and B F in the line of their major axes, will have a perfect rolling contact. Two equal logarithmic spirals have also the same property.

Let D P C be the common tangent to the point of contact P; from A and B let fall A C and B D perpendicular to D P C; then,

$$\frac{\text{angular velocity A P}}{\text{angular velocity B P}} = \frac{\text{B D}}{\text{A C}} \text{ or } \frac{\text{B P}}{\text{A P}} \ldots (1).$$

This result may be expressed in language as follows:—*The angular velocities of the wheels are inversely as the perpendiculars let fall upon the common tangent from the centres of motion.*

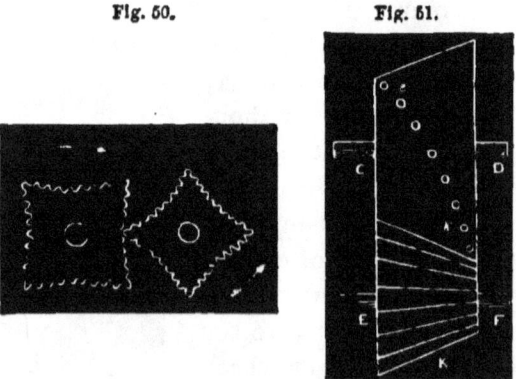

Fig. 50.   Fig. 51.

73. The form of wheels, represented in fig. 50, are used in silk-mills, and in the Cometarium.

The curves may be indefinitely varied, but they must always be constructed to answer the conditions explained in Art. 72.

74. *Roëmer's Wheels.*—E F and C D are the axes of two conical wheels or bevel-wheels K and G, having their vertices turned in opposite directions; the teeth of K are formed like those of the ordinary bevel-wheel; but the teeth on G are formed by a series of pins $e\,k$, fixed on the surface of the frustum G. By varying the relative position of these pins, any given velocity ratio may be obtained.

75. Various combinations have been invented for producing a varying angular velocity; such as the eccentric crown wheel and broad pinion, the eccentric spur-wheel with a shifting intermediate wheel, and so on.

INTERMITTENT AND RECIPROCATING MOTIONS PRODUCED BY WHEEL WORK, HAVING ROLLING CONTACT.

76. The following is an example of an intermittent motion produced by the continuous motion of a toothed wheel.

A driving wheel A, having sunk teeth on a portion of its edge, communicates an intermittent motion to the wheel B, which has a corresponding number of teeth on a portion of its edge. The portion D C of the wheel B, being a plain arc of a circle described from A as centre, allows the plain

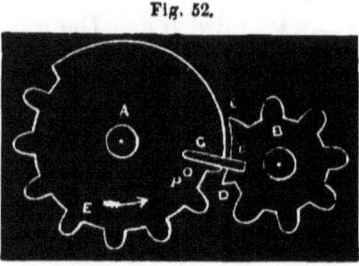

Fig. 52.

portion of the wheel A to revolve without any interruption. The wheels are brought into gear by a pin p, fixed to the wheel A, and a GUIDE-PLATE G e, fixed to the wheel B. Now when A revolves, in the direction of the arrow, the plain portion of its edge runs past D C without moving the wheel B, and at the same time keeps it from shifting; but when the pin p comes into contact with the guide-plate, the wheel B is moved round, and the teeth D E engage themselves with the teeth on B, and thus the wheel B is constrained to make a revolution; it then remains at rest until the pin p again comes round to meet the guide-plate.

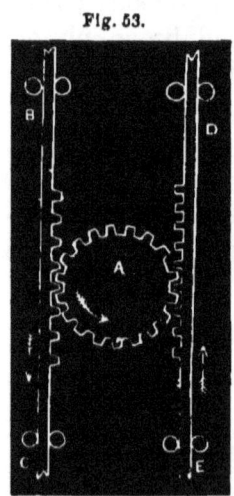

Fig. 53.

77. *The Rack and Pinion.*— By this combination a circular reciprocating motion is changed into a reciprocating rectilinear one. Teeth are cut upon the edge

of the straight bars, B C and D E, so as to work with the teeth upon the pinion A. These toothed bars are called *racks*, and they are constrained to move in rectilinear paths by guides or rollers. The racks in this combination move in opposite directions.

78. Fig. 53 represents an application of the double rack, for converting a continuous circular

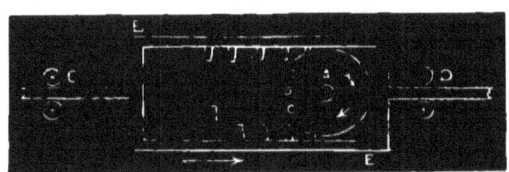

Fig. 54.

motion of a wheel, A, into a reciprocating rectilinear motion, given to the frame B E.

The teeth on A are formed by pins or staves placed about one quarter round the face of the wheel; these staves act alternately upon the racks formed on the upper and under sides of the frame. The tooth on each rack, which comes first into contact with the stave of the pinion, is made longer than the others, in order that the first stave should act obliquely upon it, thereby tending to lessen the shock. In this figure the lower stave is represented as leaving the last rack on the under side, and the upper stave as commencing its action on the elongated tooth of the upper rack.

V. ON SLIDING PIECES, PRODUCING MOTION BY SLIDING CONTACT.

The Wedge or Movable Inclined Plane.

79. Let A B C be a movable inclined plane or wedge, sliding along the smooth surface D E, by a pressure P applied to the end B C, and producing a vertical motion in a heavy rod G P₁ resting on the plane A C, and constrained to move in a straight path by means of guide rollers. The velocity ratio of P and P₁ will be constant, being expressed by the following equality:

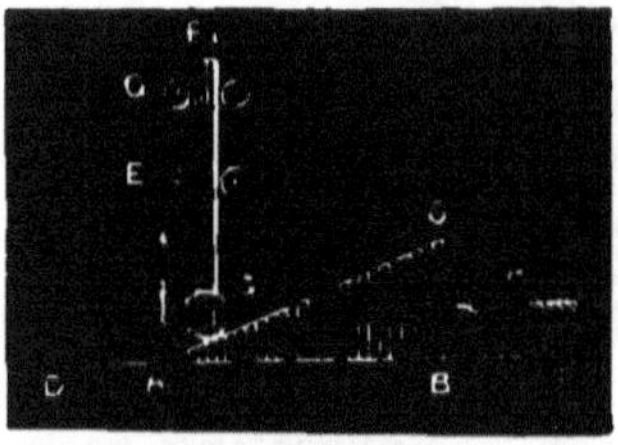

$$\frac{\text{velocity P}}{\text{velocity P}_1} = \frac{AB}{BC} \text{ or } \frac{\text{length of the wedge}}{\text{thickness of the wedge.}}$$

Fig. 55.

To transmit motion from an axis A D, *to another axis* B C, *parallel to it.*

80. The axis A D carries an arm A E, and a pin E F, which enters and slips freely in a slit made in the arm G B attached to the axis B C. When the axis A D revolves, it communicates a rotation in the same direction to the other axis B C,

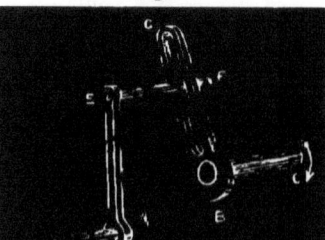

Fig. 56.

but with a varying velocity ratio, for the pin F continually changes its distance B F from the axis B C.

When the distance between the parallel axes is small, and the axis A D revolves uniformly, the angular velocity of the axis B C varies, very nearly, inversely as the distance, B F, of the pin from this axis.

The Eccentric Wheel.

81. This mechanism is usually employed to give motion to the slide-valve of the steam engine.

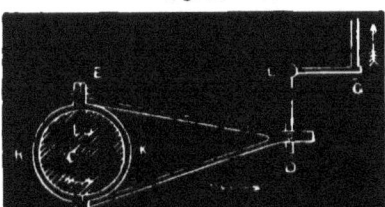

Fig. 57.

In fig. 57, B represents the axis of the eccentric wheel; C the centre of the circle; E R F K a hoop which embraces the eccentric wheel so that the one may revolve freely within the other; E F D a frame connecting this loop with the extremity D of the bent lever D L G, turning on the fixed centre L. Now when the eccentric wheel revolves in the direction of the arrow, shown in the figure, the frame with the pin D is pushed to the right, and when the lob side of the eccentric has passed the line of centres, B and D, the frame with the pin D is drawn to the left, and so on. Thus the continuous rotation of the axis B produces a reciprocating circular motion in the pin D. The stroke of the pin D will be equal to twice C B, or double the eccentricity of the wheel.

Cambs, Wipers, and Tappets.

82. Cambs are those irregular pieces of mechanism to which a rotatory motion is given for the purpose of producing, by sliding contact, reciprocating motions in rods and levers.

83. In fig. 58, B C D represents the camb, turning on its axis A, and giving a reciprocating rectilinear motion to the heavy rod E F, which is restrained to move in its rectilinear path by the guide rollers. The rotation of the axis A being in the direction of the arrow, the rod E F has an upward motion until the extreme point B of the camb comes in a line with the rod, then the portion B G of the camb allows the rod to fall, by its own weight, or by the action of a spring, until the point G comes in a line with the rod, and so on; thus one revolution of the camb, here presented, will cause the rod to make three upward and three downward strokes. By varying the curve of the camb, any law of motion may be given to the rod.

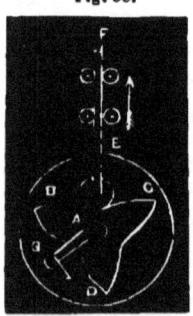

Fig. 58.

84. In fig. 59, the pin E of the rod is made to traverse a groove E G D, cut in the camb plate, so

SLIDING PIECES. 77

that the pressure of the camb upon the pin produces the downward stroke of the rod as well as its upward stroke. In this case the rod will only make one upward and one downward stroke in every revolution of the camb plate. The length of the stroke of the rod will be equal to the difference between A D and A G, where D is the point in the groove furthest from the centre A, and G is the point nearest to it.

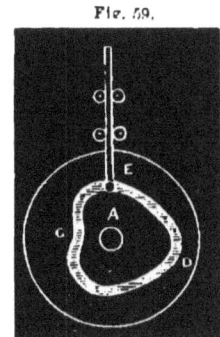

Fig. 59.

85. *To find the curve forming the groove of a camb, so that the velocity ratio of the rod and the axis of the camb may be constant.*

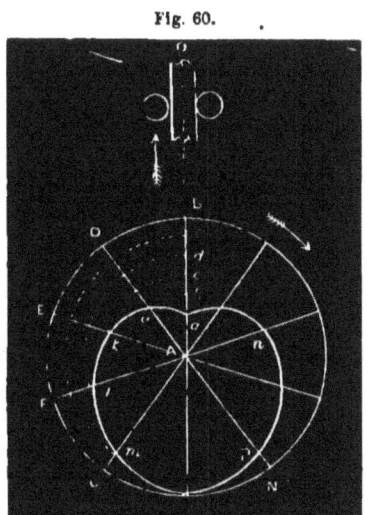

Fig. 60.

Let A be the centre of the camb, and C A B Q the direction of the rod. From A as a centre, with any convenient distance A C, describe the circle C E D B N. On B A take B *a* equal to the length of the stroke of the rod; divide it into any

7*

convenient number of equal parts, say five, in the points, $b, c, d, e,$ and divide the semicircle B D E F G into the same number of equal parts by the radial lines, A D, A E, A F. From A as a centre, with A b, A c, A d, A e, as radii, describe circles cutting A D, A E, &c., respectively in the points $g, k, l, m:$ then through these points draw the curve $a\,g\,k\,l\,m$ c; and similarly in the semicircle B N C draw the other curve $a\,n\,p$ c.

All lines drawn through the centre A of this curve are equal; thus a c $= l\,n = g\,p =$ &c. Hence if the rod had two pins placed at a and c, the camb would revolve between them, and would cause the rod to make a downward as well as an upward stroke. This curve is the spiral of Archimedes.

By dividing the line B a into parts having a

Fig. 61.

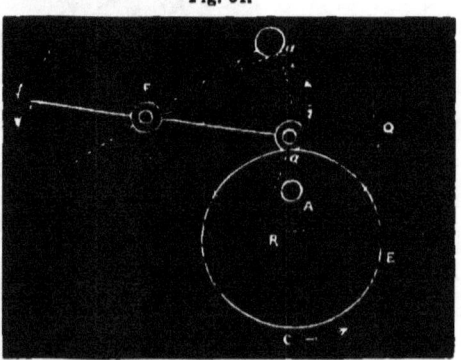

varying ratio to one another, any proposed law of velocity may be given to the rod.

SLIDING PIECES.   79

86. In fig. 61, the continuous rotation of the camb A E C, revolving on the axis A, gives an oscillating motion to the rod or lever F a, turning on the centre F. In one revolution of the camb the rod makes a double oscillation in the arc $a\ a_1$.

87. *Wipers.*—When the rod is to receive a series of lifts with intervals of rest, the camb is made into the form of projecting teeth, which are commonly called *Wipers* or *Tappets*.

Fig. 62.

88. In fig. 62, the revolving cylinder C has five wipers upon its circumference, which give five downward strokes to the hammer, H, placed at the

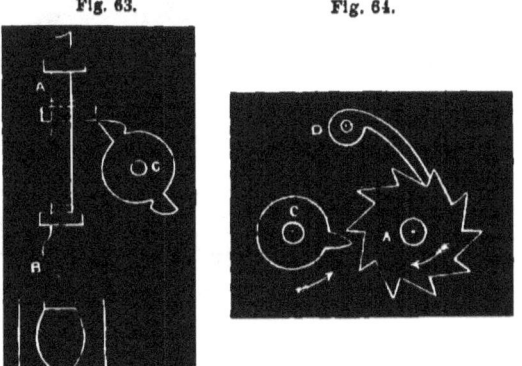

Fig. 63.   Fig. 64.

extremity of the lever A H, in each revolution of the cylinder.

89. In fig. 63, two tappets, upon the revolving cylinder c, give two downward strokes to the heavy bar or *stamper* A B, in each revolution of the cylinder. In this case the bar A B is constrained to move in a rectilinear path by means of guide rollers.

90. In fig. 64, a single wiper on the cylinder c gives an intermittent rotation to the ratchet wheel A with its detent D. At each revolution of c only one tooth in A is moved round, so that for the greater portion of the revolution of c the wheel B is at rest.

91. In fig. 65, the continuous rotation of three

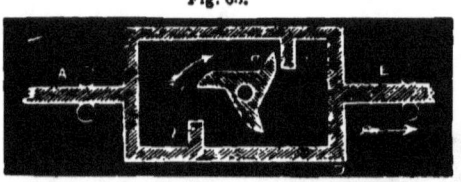

Fig. 65.

wipers a, b, c, communicates a reciprocating rectilinear motion to the frame A B C D. The wiper a is engaged with the *pallet* e, and at the instant of disengagement the wiper b becomes engaged with the pallet g, and then the frame starts its motion in a direction contrary to that of the arrows; and so on.

The Swash Plate.

92. By this mechanism, the continuous rotation of an axis produces a reciprocating recti-

SLIDING PIECES.   81

linear motion in a rod, in the direction of its length.

Here C E represents the revolving axis, to the top of which is fixed the inclined circular flat plate A B, called the swash plate; A D F the rod to which a reciprocating motion is given in the the direction of its length, having a frictional wheel A at its lower extremity resting on the swash plate. This rod is kept in contact with the plate by its own weight, or, if this be not sufficient, by means of a spring. Now as the swash plate turns round, the rod A F is alternately raised and depressed, so that at every revolution of the plate the rod performs an upward and a downward stroke. Supposing the rod, as represented in this figure, to be at the lowest point of its stroke; from C, the centre of motion of the plate, let fall C D perpendicular to A F; then A D will be equal to half the stroke of the rod. Moreover, let θ be any angle moved over by the axis, and let h be the corresponding space moved over by the extremity A of the rod; then

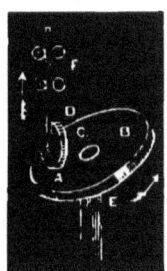

Fig. 66.

$$h = A D + (1 - \cos \theta),$$

which gives the position of the rod at any point of the rotation of the plate.

93. There are an almost endless variety of com-

binations for producing reciprocating motions of this kind, by means of sliding contact.

94. In fig. 67, an eccentric revolving pin *e*, sliding or working in the slit of the arm *r s* gives a reciprocating motion to the rod *p q* in the direction of its length.

95. In fig. 68, the same effect is produced by

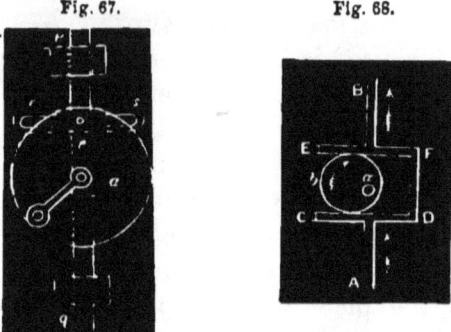

Fig. 67.   Fig. 68.

the rotation of an eccentric wheel, *a b*, on its axis *a*, within the frame C D F E.

SCREWS.

96. *Construction of a Helix or Screw.*—Let A *a* K be a cylinder, and A D E a piece of paper cut in the

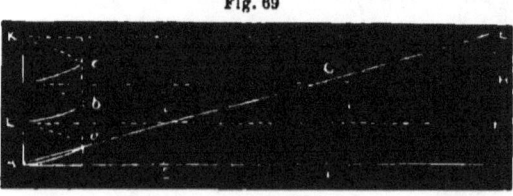

Fig. 69

form of a right angled triangle, having its height D E equal to the height A K of the cylinder. Now if this paper be wrapped round the cylinder, the slant edge A E of the paper will trace the helix or screw A *a* L *b c* K upon the cylinder. If A B = B C = C D be equal to the circumference of the cylinder, the edge of the paper will form four convolutions, and the perpendiculars B F = I G = H E will be the distance between the threads of the screw.

97. *The pitch of a Screw* is the distance B F between two successive convolutions. If $t =$ B F, the distance between the threads of the screw, $r =$ the radius of the cylinder, $\theta =$ angle B A F; then

$$t = \frac{2 \pi r}{\tan \theta}.$$

98. We may also conceive the helix of the screw to be formed by the compound motion of a point. Suppose the cylinder to rotate uniformly upon its axis, whilst a point A upon its surface at the same time moves uniformly in the direction of its length: then, with this compound motion, the point A will trace the helix of a screw.

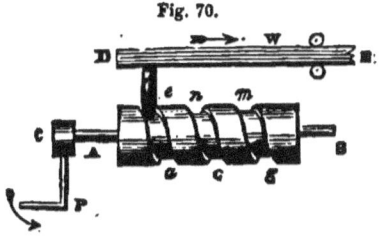

Fig. 70.

99. *Transmission of motion by the screw.*—Let *e a n c m g* be

a spiral groove cut upon a cylinder; A B the axis on which it turns; D E a rod parallel to the axis A B, and constrained to move in the direction of its length; *e* a tooth attached to this rod fitting the groove of the screw. Now when the wheel C is turned in the direction of the arrow, the tooth with the rod D E will be moved from left to right in the direction of its length, that is, parallel to the axis of the screw.

The velocity ratio of the wheel C and the rod D E will be constant, for we have

$$\frac{\text{velocity c}}{\text{velocity c d}} = \frac{\text{circum. described by c}}{\text{pitch of the screw}}.$$

If R be the radius of the wheel C, $r=$ the radius cylinder A B, V $=$ velocity circum. C, v the velocity of the bar D E, and so on as in art. 97; then the above equality becomes—

$$\frac{V}{v} = \frac{2\pi R}{t} \ldots (1);$$

$$\therefore \frac{V}{v} = \frac{R}{r} \tan \theta \ldots (2);$$

and when R $= r$, then—

$$\frac{V}{v} = \tan \theta \ldots (3);$$

that is, *the velocity ratio is equal to the tangent of the angle which the thread of the screw makes with the sides of the cylinder.*

100. It is obvious that the number of teeth in the bar D E will not at all alter its motion.

In fig. 71, the screw acts upon a series of teeth

Fig. 71.

upon the *rack* D E. This arrangement, called *the rack and screw*, converts a circular motion into a rectilinear one.

Solid Screw and Nut.

101. In general the piece acted upon by the screw has its teeth, or rather its threads, formed in a cavity which embraces the whole circumferences of the screw, and the threads of the one exactly fitting the threads of the other. This modification is shown in fig. 72, where N is the hollow screw fitting the threads of the screw S. The solid piece S is called the male screw; and the hollow piece the female screw or *nut*.

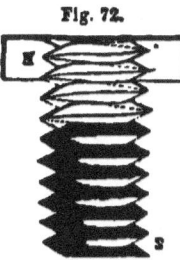
Fig. 72.

102. Screws are either left-handed, or right-handed, according to the direction of the threads.

103. It is important to observe that the following relations of motion subsist between the solid screw and the nut:

1. When the nut is fixed, the solid screw will have a motion in the direction of its length, upon being turned round.

2. If the nut revolves, without having any longitudinal motion, the solid screw will have a motion in the direction of its length, provided it is incapable of revolving.

3. If the solid screw revolves without having any motion in the direction of its length, the nut will have a longitudinal motion, provided it is incapable of revolving.

The first two cases are exemplified in the different forms which are given to the common press, and the last case is exemplified in the construction of the self-acting slide rest of the lathe, and in other kinds of mechanism.

The screw is usually employed for producing very slow uniform motions, and for exerting great pressure through a limited space.

The Common Press.

104. In fig. 73, s s is the solid screw, n the nut, n p the lever, b the lower press board which is constrained to move in an upward direction by means of the guide frame.

Case 1. In this case the nut n revolves, but does not move longitudinally, but the screw s s is incapable of revolving. Hence the press board b is moved upward at every revolution of the nut, over a space equal to the pitch of the screw, or the distance between the threads, that is,

$$\frac{\text{velo. P}}{\text{velo. B}} = \frac{\text{circum. described by P}}{\text{distance between the threads.}}$$

Example.—Let the distance between the threads $= \frac{1}{4}$ in., the length of the lever N P $= 2\frac{1}{2}$ ft.; required the velocity ratio of the point P and the press-board B.

$$\frac{\text{velo. P}}{\text{velo. B}} = \frac{2 \times 2\frac{1}{2} \times 12 \times 3\cdot1416}{\frac{1}{4}} = 753\cdot984.$$

That is, the velocity of P is 753·984 times that of B.

Case 2. In this case N is a perforated cylinder forming part of the solid screw S S, and therefore turns with it on a pivot which works in a socket placed on the under side of the press board B; the piece K fixed to the frame contains the hollow or female screw; so that the solid screw, S S, is capable of revolving and of moving longitudinally, whilst the nut K remains absolutely fixed.

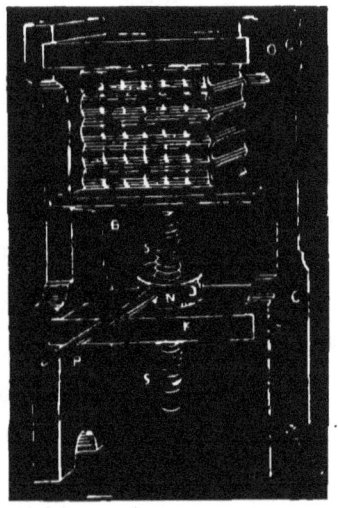

Fig. 73.

Compound Screw.

105. This mechanism consists of two screws A and D, the smaller one D working within the larger one A. The screw A works in a fixed nut or female screw at K, and is capable of revolving and moving in the direction of its length; the small screw D is incapable of revolving, but is capable of moving in the direction of its length. In one revolution of the lever P, the screw A descends a space equal to the distance between its threads,

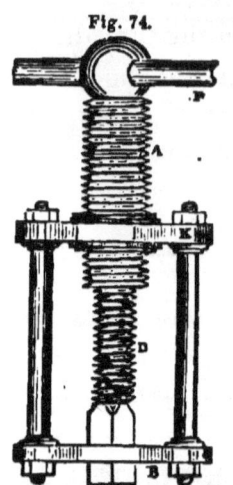

Fig. 74.

but at the same time the screw D enters the hollow screw formed in A, a space equal to the distance between the threads on D, so that the extremity B will only descend a space equal to the difference between the thickness of the threads on A and the thickness of the threads on B; hence we have

$$\frac{\text{velo. P}}{\text{velo. B}} = \frac{\text{circum. described by P}}{\text{dist. bet. th'ds on A} - \text{dist. bet. th'ds on D.}}$$

If the length of the lever $P = r$, the pitch of the screw $A = t$, and the pitch of $D = t_1$; then

$$\frac{\text{velo. P}}{\text{velo. B}} = \frac{2 \pi r}{t - t_1} \ldots (1).$$

Example.—Let $r = 5$ ft., $t = \frac{1}{2}$ in., $t_1 = \frac{3}{8}$ in.; then

$$\frac{\text{velo. P}}{\text{velo. B}} = \frac{2 \times 5 \times 12 \times 3\cdot1416}{\frac{1}{2} - \frac{3}{8}} = 3015\cdot936$$

The same velocity ratio might be attained by making the pitch of a single screw A, equal to $t - t_1$, but the threads, in this case, might be too weak to stand the pressure; hence the advantage of the compound screw.

The Endless Screw.

106. When the threads or teeth of a revolving screw are made to act upon the teeth of a wheel, as in fig. 75, the mechanism is called the endless screw. Here, each rotation of the axis A B of the screw turns round one tooth of the wheel C, the pitch of the screw on the axis A B being equal to the pitch of the teeth on the wheel.

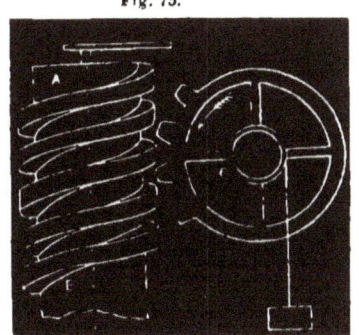

Fig. 75.

If Q and q be the synchronal rotations of the wheels and the screw respectively, and N the number of teeth in the wheel; then

$$\frac{q}{Q} = N \ldots (1).$$

If $N = 40$, then $\frac{q}{Q} = 40$; that is, for every revolution performed by the wheel the screw will make 40.

If R, r be the respective pitch-radii of the wheel and screw, θ being, as before, the angle which the thread of the screw makes with its axis; then

$$\frac{q}{Q} = \frac{R}{r} \tan \theta \ldots (2).$$

The Differential Screw.

107. A D is an axis on which are formed two screws, A B and B C, whose pitches are different.

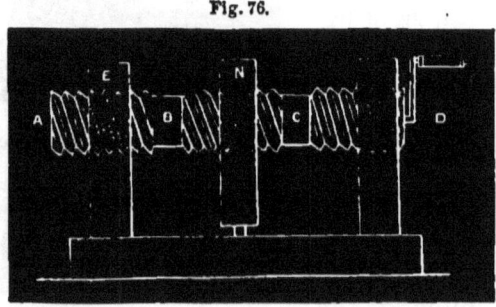

Fig. 76.

The screw A B passes through a fixed-nut or female screw E, whilst B C passes through a nut N which is capable of moving longitudinally, but incapable of revolving from the intervention of the guides.

Let the screw make one turn so as to move the cylinder from right to left, then the screw A B will move through the fixed nut E a space equal to the distance between its threads; but, at the same time, the screw B C will move through the nut N a space equal to the thickness of the threads on B C; so that the nut N will only be moved through a space equal to the difference between the thickness of the threads on A B and B C, that is—

In one revolution of A, the space moved over by the nut N = pitch screw A B — pitch screw $BC = t - t_1$, where t is put for the pitch of the screw A B, and t_1 for that of B C.

If $t = t_1$, then nut N will remain at rest.

If the screw A B be right-handed, and B C left-handed; then $t + t_1$ will be the space moved over by the nut N in one revolution of A.

The Archimedian Screw Creeper.

108. This machine is used for conveying corn

Fig. 77.

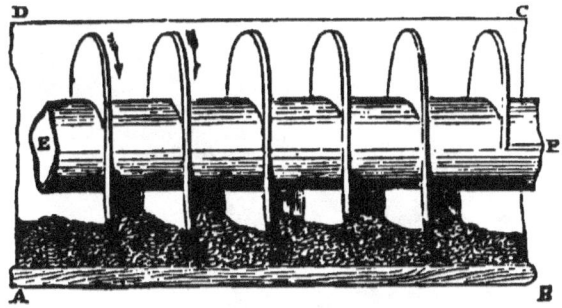

from one part of a corn mill to another. It consists of a wooden cylindrical trough, A B C D, within which revolves a shaft, E F, having a deep spiral thread formed upon its surface. The corn is dropped in at one extremity of the trough by a hopper, and by the revolution of the creeper the corn is pushed along toward the other extremity of the trough.

Mechanism for cutting Screws.

109. C D is the cylinder, or axis on which the

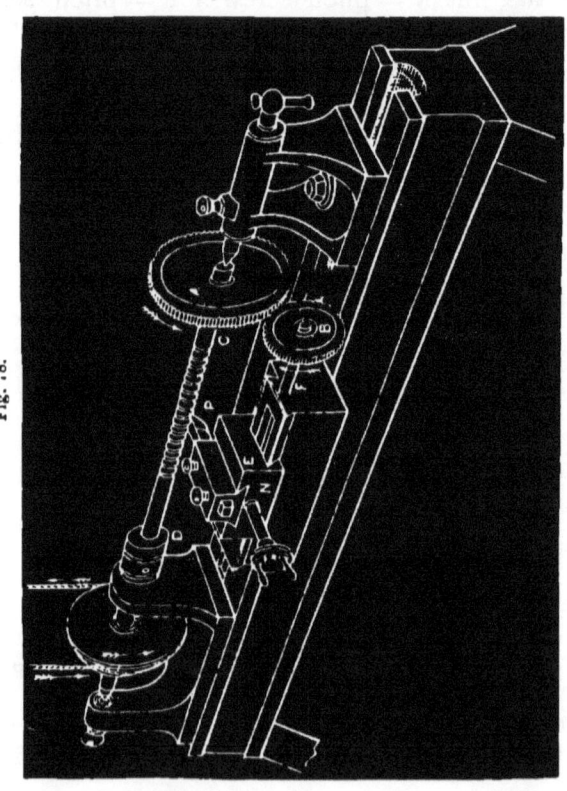

Fig. 78.

screw is to be cut, revolving with the mandril D of the lathe; A a toothed wheel revolving with the axis C D, and giving motion to the toothed wheel B, round its axis F E, on which is cut the parent screw; this screw gives a longitudinal motion to the nut N, as in Case 3, carrying the sliding table or saddle upon which is securely clamped the cutting tool P intended to cut the thread of the screw on the cylinder C D. In the place of the wheels A B, any combination of wheels may be used so as to produce any relative longitudinal velocity to the cutting tool P, and thereby to form a screw of any given pitch on C D with the same parent screw F E.

Let n = the no. of teeth on the wheel A, n_1 = the no. of teeth on B, t = the pitch of the screw on C D, t_1 = the pitch of the screw on F E; then

$$t = \frac{n}{n_1} \cdot t_1 \ldots (1),$$

which expresses the pitch of the screw on C D.

From this equality we get,

$$\frac{t}{t_1} = \frac{n}{n_1} \ldots (2),$$

that is to say, *the pitches of the screws are in the ratio of the number of teeth on their respective wheels.*

If n_1 and t_1 be constant, then

$$t \infty n,$$

that is to say, *the pitch of the screw on C D varies with the number of teeth on its wheel A.*

Let k and k_1 be the number of threads per inch on the cylinders C D and F E respectively, then

$$\frac{1}{k} = t, \text{ and } \frac{1}{k_1} = t_1,$$

and eq. (2) becomes—

$$\frac{k}{k_1} = \frac{n_1}{n} \ldots (3).$$

Now, let there be an intermediate pinion and wheel, turning on the same axis, placed between A and B; and let the pinion (acted upon by A) contain e_1 teeth, and the wheel e teeth; then the velocity ratio of the axis F E will be increased by the ratio $\frac{e}{e_1}$, and hence eq. (3) becomes—

$$\frac{k}{k_1} = \frac{n_1}{n}\frac{e}{e_1} \ldots (4).$$

Example.—Let $n = 30$, $n_1 = 10$, $t_1 = \frac{1}{2}$ in.; required t.

Here by eq. (1), $t = \frac{n}{n_1} \cdot t_1 = \frac{30}{10} \times \frac{1}{2} = 1\frac{1}{2}$ in.

To produce a changing reciprocating rectilinear motion by a combination of the camb and screw.

110. E F is a conical shaped camb, turning on the eccentric axis A B, on which is cut the screw K B, working in the fixed nut or hollow screw N; D C a rod resting on the camb, constrained to

move in the direction of its length, and to which the varying reciprocating motion is to be given. Here, whilst the camb revolves, it has a continuous motion in the direction of the axis A B, so that the lower extremity, C, of the rod D C describes a spiral or screw curve upon the cone whose pitch is equal to the pitch of the screw K B. The effect of this is to make C D reciprocate in its path in such a manner that the stroke in one direction is shorter than that in the opposite direction.

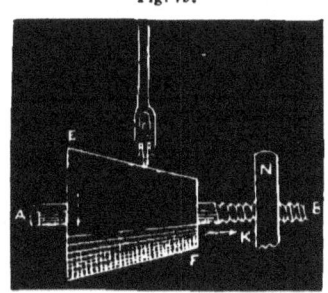

Fig. 79.

To produce a boring motion by a combination of the screw and toothed wheels.

111. Here it is required to produce a rapid rotation combined with a very slow motion in the direction of the axis.

The screw I B is cut upon a portion of the revolving axis A B; this screw passes through a nut K capable of revolving with the wheel G, but incapable of moving in the direction of its axis, as in Case 2, page 87; the wheel G is driven by the pinion F revolving on the parallel axis D C; E is a long pinion, turning on this axis, and acting on the wheel D, which transmits a rotatory motion to the screw axis A B. Now the rotation of C D produces a rotatory motion in the axis A B, and at the same time causes it to advance, in the direc-

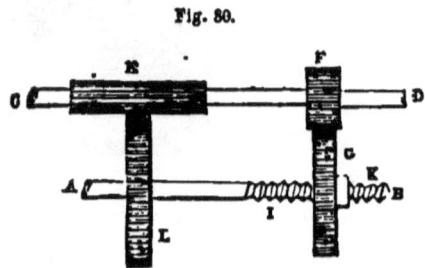

Fig. 80.

tion of its length, with a velocity determined by the following formula.

Let Q, Q_1, q_1 be the synchronal rotations of the axis C D, the nut K and wheel G, and the wheel and axis A B, respectively; N, N_1, n, n_1, the number of teeth in the wheels F, G, E, L, respectively; s the space moved over by A B in the direction of its length, and $t =$ the pitch of the screw I B.

Now Q_1 rotations of the nut K moves the screw A B through a space equal to $Q_1 \times t$; but q_1 rotations of L moves the screw through a space, in the *opposite* direction, equal to $q_1 \times t$; therefore in Q rotations of the axis C D, the screw A B will be moved through a space equal to the difference between $Q \times t$ and $q_1 \times t$, that is,

$$s = (q_1 - Q_1) t;$$

but $\dfrac{Q_1}{Q} = \dfrac{N}{N_1}$, and $\dfrac{q_1}{Q} = \dfrac{n}{n_1}$;

$$\therefore s = \left(\dfrac{n}{n_1} - \dfrac{N}{N_1}\right) Q\, t \ldots (1).$$

Now the difference $\dfrac{n}{n_1} - \dfrac{N}{N_1}$ may be very small as compared with Q, and consequently s may be made as small as we please as compared with Q, which is the condition required for the construction of a boring instrument. The boring tool is placed upon one extremity of the axis A B.

MACHINERY OF TRANSMISSION.

CHAPTER II.

ON WHEELS AND PULLEYS.

THE elementary principles of motion by rolling contact and by wrapping connectors have already been explained, so that in the present chapter we have only to examine in detail the methods of applying these principles and their respective advantages, and especially the mode of constructing wheels in gear, so that the resulting motion shall most nearly approach the condition of perfect rolling contact.

We saw in the preliminary chapter that there were two methods of transmitting power through trains of wheel work, the first being through the agency of wrapping connectors, and the second by rolling contact.

Wrapping connectors.—Considerable difference of opinion exists as to the best and most effective principle of conveying motion from the source of power to the machinery of a mill. The Americans prefer leather straps,* and large pulleys or riggers.

---

* I have selected the word *strap*, instead of *belts* or *bands*, as a term more generally applied to wrapping connectors in the northern districts.

In this country, and especially in the manufacturing districts, toothed wheels are almost universally employed. In some parts of the South, and in London, straps are extensively used; but in Lancashire and Yorkshire, where mill-work is carried out on a far larger scale, gearing and light shafts at high velocities have the preference. Naturally, I am of the opinion that the North is right in this matter, and that consistently, as I was to a great extent the first to introduce that new system of gearing which is now general throughout the country, and to which I have never heard any serious objection. I have been convinced by a long experience that there is less loss of power through the friction of the journals, in the case of geared wheel work, than when straps are employed for the transmission of motive power. Carefully conducted experiments confirm this view, and it is therefore evident which mode of transmission is, as a general rule, to be preferred.

There are certain cases in which it is more convenient to use straps instead of gearing. With small engines driving saw-mills, and some other machinery where the action is irregular, the strap is superior to wheel work, because it lessens the shocks incidental to these descriptions of work. So, also, when the motive power has been conveyed by wheel work and shafting to the various floors of a mill, it is best distributed to the machines by means of straps.

In some of the American cotton factories, however, there is an immense drum on the first motion, with belts or straps from two to three feet wide, transmitting the power to various lines of shafting, and these in turn through other pulleys and straps, giving motion to the machinery. From this description it will be seen that the whole of the mill is driven by straps alone, without the intervention of gearing.

The advantages of straps are, the smoothness and noiselessness of the motion. Their disadvantages are, cumbrousness, the expense of their renewal, and the necessity for frequent repairs. They are inapplicable in cases where the motion must be transmitted in a constant ratio, because, as the straps wear slack, they tend to slip over the pulleys, and thus lose time. In other cases, as has been observed, this slipping becomes an advantage, as it reduces the shock of sudden strains, and lessens the danger of breaking the machinery.

Very various materials are employed for straps, the most serviceable of all being leather spliced with thongs of hide or cement. Gutta percha has been employed with the advantage of dispensing with joints, but it is affected by changes of temperature, and it stretches under great strains. Flat straps are almost universally employed, in consequence of the property they possess of maintaining their position on pulleys, the edge of which is slightly convex (fig. 81). Round belts of catgut or hemp are sometimes used, running in grooves, which are better made of a triangular than a circular section—so that the belt touches the pulley in two lines only, tangential to the sides of the groove; in this case the friction of the belt is increased in proportion to the decrease of the angle of the groove.

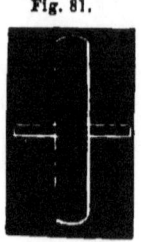

Fig. 81.

The strength of straps must be determined by the work they have to transmit. Let a strap transmit a force of n horses' power at a velocity of v feet per minute, then the tension on the driving side of the belt is $\dfrac{33000\,n}{v}$ lbs. independent of the initial tension producing adhesion between the belt and pulley. For example, let v be 314·16 feet per minute, or the velocity of a 24-inch pulley at 50 revolutions per minute, and let 3 horses' power be transmitted; then $\dfrac{33000 \times 3}{314\cdot16} = 312$ lbs., the strain on the pulley due to the force transmitted.

The following table has been given for determining the least width of straps for transmitting various amounts of work over different pulleys. The velocity of the belt is assumed to be between 25 and 30 feet per second, and the widths of the belts are given in inches. With greater velocities the breadth may be proportionably decreased.

TABLE I.—APPROXIMATE WIDTHS OF LEATHER STRAPS, IN INCHES, NECESSARY TO TRANSMIT ANY NUMBER OF HORSES' POWER.

Horses' Power.	Smallest Diameter of Pulley in Feet.							
	1	2	3	4	5	6	7	8	10
1	3·6	1·8	1·2	—	—	—	—	—	—
2	7·2	3·6	2·4	1·8	1·4	—	—	—	—
3	10·8	5·4	3·6	2·7	2·1	1 8	1·5	—	—
4	14·4	7·2	4·8	3·6	4·8	2·4	2·0	1·8	1·4
5	18·0	9·0	6 0	4·5	3·6	3·0	2·5	2 2	1·8
7	25·2	12·6	8·4	6·3	5·4	4·2	3·5	3·7	2·5
10	36·0	18 0	12·0	9·0	7·2	6·0	5·1	4·5	3·6
12	43·2	21·6	14·4	10·8	8·6	7·2	6·1	5·4	4·3
14	—	25·2	16·8	12·6	10·0	8·4	7·1	6 3	5·0
16	—	28·8	19·2	14·4	11·5	9·6	8·2	7·2	5·7
18	—	32·4	21·6	16·2	12·9	10·8	9·2	8·1	6·4
20	—	36·0	24·0	18·0	14·4	12 0	10·2	9 0	7·2
25	—	45·0	30·0	22·5	18·0	15·0	12·8	11·2	9·0
30	—	—	36·0	27·0	21·0	18·0	15·0	13·0	10·0
40	—	—	48·0	36·0	28·0	24·0	20·0	18·0	14·0
50	—	—	—	45 0	36·0	30·0	25·0	22·0	18·0
60	—	—	—	—	43·0	36·0	30·0	27·0	21·0
70	—	—	—	—	—	42·0	35·0	31·0	25·0
80	—	—	—	—	—	—	41·0	36·0	28·0
100	—	—	—	—	—	—	51·0	45·0	36·0

Toothed Wheels.—The second method of communicating motion is by rolling contact, as explained in the preliminary chapter.* But, in practice, the adhesion between the surfaces is seldom sufficient to communicate the necessary power, and hence various contrivances—such as the wheel and trundle; and toothed wheels—have been substituted. The general equations for velocity, ratio, etc., are the same as if the wheels rolled on each other at the pitch circles, but in fact each tooth slides upon its fellow. The determination

---

* See page 56.

of the best forms of these teeth so that the friction shall be a minimum and the motion uniform, is one of the most important contributions of applied mathematics to practical engineering.

Of the introduction of toothed wheels and toothed gearing, we know very little. Hero of Alexandria, who wrote two centuries before our era, speaks of toothed wheels and toothed bars in a way which seems to indicate that he was not altogether ignorant of this method of transmitting motion. Later forms are figured in great variety in the different collections of mechanical appliances of the sixteenth and seventeenth centuries.

Spur gearing is employed where the axes on which the wheels are placed are parallel to one another. The smaller wheel in a combination of this sort is termed the pinion. Annexed (fig. 82) is a pinion from Ramelli (A. D. 1588), which from its form, may be surmised to be of metal. The principle on which spur gearing is constructed is primarily the communication of motion through the rolling of two cylinders on one another. The teeth are introduced to prevent slipping, and thus to insure the regular communication of the motive power.

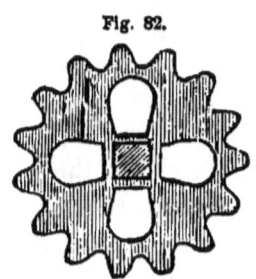

Fig. 82.

In the older wooden wheels, the teeth were usually formed of hard wood, and driven into mortises on the periphery of a wooden wheel. The pinions were generally replaced by trundles, in which cylindrical staves, fixed at equal distances round the periphery of two discs, were driven by the teeth of the wheel.

The mortise wheels are still retained in countries where iron is expensive, and even in this country they are employed in a modified form. Iron pinions, with wooden cogs fixed in the periphery, are used to receive the motion from the fly-wheels of engines, with a view to reduce the noise and to increase the smothness of the motion; and many millwrights prefer, in all cases where large wheels are required to run at high velocities, to make one of them a mortise-wheel, with wooden cogs.

There does not appear to have been much improvement in the construction of wood and iron gear since it was first introduced by Mr. Rennie; the only exception being the introduction of a machine for cutting out the form of the teeth,* which in those days was done by hand, with keys or wooden wedges fitting into dovetails in the 'shanks' of the cogs, as shown at a, fig. 83, on the

---

* Mr. Smiles states, in his ' Lives of the Engineers,' that Brindley, more than a century ago, invented machinery for the manufacture of tooth and pinion wheels, ' a thing,' as stated by the author, ' that had not before been attempted, all such wheels having, until then, been cut by hand, at great labor and cost.'

concave side of the rim; now they are made with an iron pin driven through the cog, close to the rim, as at *b*. The iron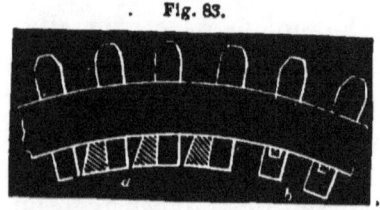

Fig. 83.

pinion or wheel intended to work in contact with the wood teeth was, up to a recent date, turned and carefully divided to the epicycloidal form,

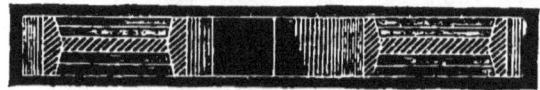

Fig. 84.

Fig. 85.

and then chipped and filed with great exactitude, in order to fit accurately into the wooden teeth of the driving wheel. In all the corn mills of the present day, and where great speed is required, the same attention to accuracy is observed in wood and iron gear as in former times.

The greatest advance in the application of gearing resulted from the introduction, at the end of the eighteenth century, of cast iron in place of wood. The credit of the introduction of this material is usually given to Smeaton, who began to use cast iron in the construction of the Carron Rolling Mill, in 1769. But the late Mr. John Rennie, when at Boulton and Watt's, in 1784, was probably the first to carry the use of cast iron into all the details of mill work. Figs. 84, 85 are copied from the original designs for the Soho Rolling Mill, dated 1785. But the Albion Corn Mills, built about the same time (1784–5), may be considered as the earliest instance of the entire replacing of wood by cast iron for the bevel and spur wheels and shafts. This was effected by the same distinguished engineer.

Fig. 86.

Where the shafts of the wheel and pinion are

not parallel to each other, various forms of conical trundles and bevel wheels, are employed. The simplest plan is probably the face wheel and trundle shown in fig. 86, which have been employed from a very early period, and which, if made of metal, take the form of the crown wheel and pinion, fig. 87. Where the axes are not at right angles, conical trundles have been used, one of which is figured in Bessoni (A. D. 1578.)

The most perfect arrangement, however, is that in which two wheels called bevel wheels are employed constructed in the form of frusta of cones. These were not introduced till the middle of the last century, the principles of the construction of the teeth being due to Camus (A.D. 1752).

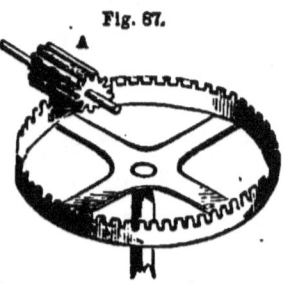

Fig. 87.

Fig. 88 shows a bevel wheel designed for the Rolling Mill at Soho, by the late Mr. Rennie, in 1785.*

---

* It is evident from the shape of the eye of these wheels, figs. 84, 85, and 88, that they were intended for wooden shafts, and that cast iron had not been in use much before that time. At an earlier period, Mr. W. Murdock, of Soho, had a cast iron bevel wheel, which was considered the first introduced into Scotland, many years previous to the above date. Mr. Smeaton also had introduced iron wheels at Carron in 1754, and afterwards at a mill at Belper, in Derbyshire. (See Smiles's "Life of Rennie," page 138.)

The smoothness and economy of wheel work depend entirely upon the accuracy of the curvature of the individual teeth which gear with one another. Two chief defects result from imperfections in their construction: first, the motion com-

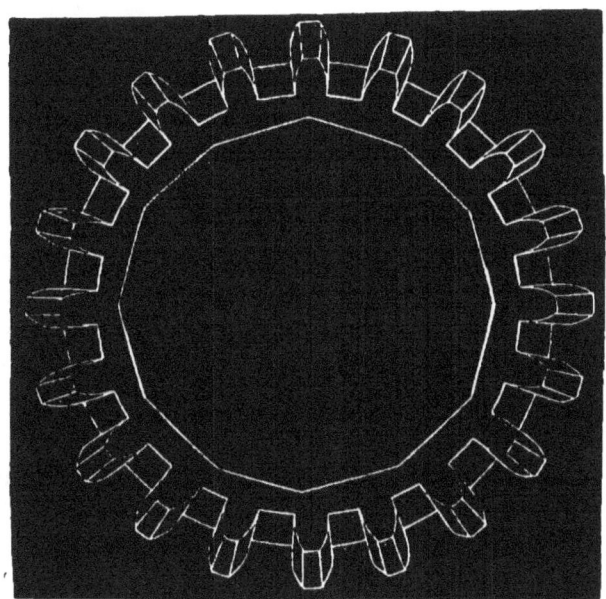

Fig. 88.

municated to the driven wheel is irregular, increasing and diminishing alternately as each tooth passes the line of centres; and, second, there is an unnecessary friction between the teeth in gear, resulting not only in loss of power, but also causing a great and destructive wear in the teeth and journals. These defects can only be avoided by reducing, as far as praticable, the size of the teeth, and by the adoption of true principles in setting out their curvature in the original model.

To the first cause alone a large part of the perfect action of modern machinery of transmission is to be attributed; but there is moreover no doubt that, in practice, even where true principles have not been adopted, a considerable approach has been made to such forms as theory requires. Now, with certain limitations, it is known that if any form of tooth be taken for one wheel, there can be found another tooth which will work correctly with it. But there are certain forms which, being susceptible of accurate mathematical determination, are most convenient for the purpose. Camus, in 1752, was the first to work out the properties of epicycloidal and hypocycloidal curves when employed in the construction of the teeth of spur and bevel gearing. De la Hire adopted the same form. Euler, in 1760, and Kaestner, in 1771, investigated in a similar manner the properties of the involute. Since their time, Ferguson, Buchanan, Hawkins, Rennie, and Airy, have all contributed to perfecting the mathematical theory. And Professor Willis, amongst other important additions, has shown how a close approximation to a true form may be made by the adoption of a system of circular arcs.

From 1788, when Rennie completed the Albion

Mills, to the present time, wood and iron gear have been in general use for high velocities, and for every description of machinery where smoothness and accuracy of motion were required. Mr. Rennie was the first to introduce this system; and in most cases he made the driver, or large wheel, with wood cogs, and the driven, or pinion, of iron "chipped and turned"—that is, every tooth of the iron wheel was carefully divided in the pitch, having first been turned on the fane and the ends of the teeth, and drawn to the epicycloidal form. They were then chipped with the hammer and chisel, and accurately filed to the required dimensions and forms. The same process was applied to the wooden teeth; and these wheels, when duly prepared, were keyed on their respective shafts, and securely fixed in contact in the mill.

The chipping and filing process has of late years been superseded by a cutting machine, which effects the same purpose, with less risk of error; and *the good old system of a penny an inch*, as practised in Rennie's time, has been exploded, much to the discomfiture of the old millwrights, who adhere with great tenacity to the hammer and chisel. Fig. 89 shows the cutting machine as constructed by Messrs. Peter Fairbairn and Co., of Leeds.

The object of this machine is not only to pitch and trim the teeth of a large spur or other wheel,

112   MACHINERY OF TRANSMISSION.

Fig. 89.

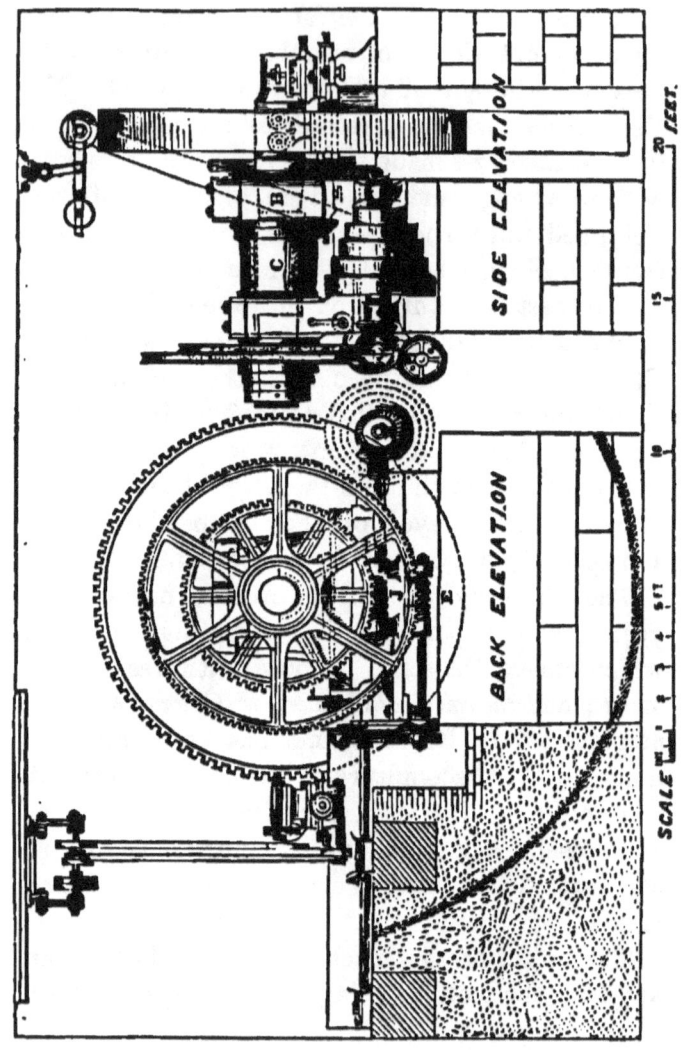

but to turn the face and sides of the segments previously, when bolted to the arms.

When used as a lathe for turning, the parts in use are as follows: B is a large headstock, carrying a hollow spindle (C), through which is inserted a mandrill upon which the wheel to be cut and turned is keyed. Provision is made for carrying the other end of this mandrill by a loose fixing. The hollow spindle is driven (with the wheel upon it) by a worm wheel (J) which is made to run loose on the spindle, but which is now by a lock bolt connected to the larger worm wheel or dividing wheel (E), the worm of which is now thrown out, and which is keyed firmly on the spindle. The necessary speeds are given by the five-speed cone and mitre gear. The tool for turning is held in an ordinary slide rest, which moves transversally on a saddle, which slides and is fastened in the T groves of two strong beds (A), firmly secured to masonry, and between which the wheel revolves.

When used for pitching and trimming, the lock bolt connecting the two worm wheels is removed, and the pitch is given by the train of change wheels and division plate (A). The place of the slide rest is now taken by a headstock carrying two cutters, one for roughing, and the other for finishing.

The finishing rose-cutter is the counterpart of
10*

the space between the teeth, and is transversed across, making both sides of the tooth alike.

The remainder of the arrangement will be obvious from the sketch. The same machine can be also readily arranged for cutting worm-wheel teeth, or for bevel gear.

The best form which can be given to the teeth of wheels is that which will cause them to be always, in regard to the power they mutually exert, in equally favorable situations, and, consequently, will give the machine the property of being moved uniformly by a power constantly equal. This would be accomplished by simple rolling contact, which corresponds with the case in which the teeth are infinitely small.

Definitions.

1. Spur gearing is that in which the pitch lines of the driving and driven wheel are in the same plane (fig. 90).

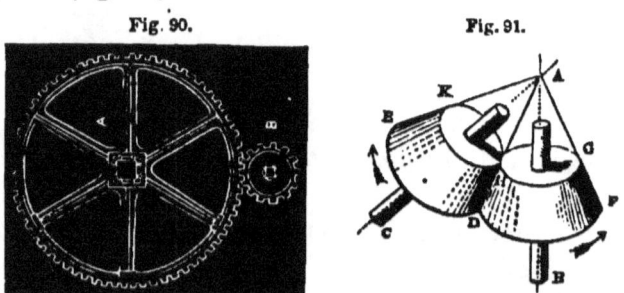

Fig. 90.   Fig. 91.

2. Bevel gearing is that in which the planes of the pitch lines of the driving and driven wheel

are inclined to each other. In practice, they are in most cases at right angles (fig. 91).

3. Of two wheels in gear, the lesser is called the pinion.

4. When two wheels are in gear, a straight line joining their centres is called the line of centres.

5. If the line of centres be divided into two parts, proportionally to the number of teeth in the wheel and pinion, these parts are called the proportional or primitive radii of the wheel and pinion.

6. The radii of the circles which limit the extremities of the teeth are called the true radii.

7. If, from the centres of the wheel and pinion, circles be drawn with radii equal to the primitive radii, so that they touch one another in the line of centres, the circles are called the pitch lines of the wheel and pinion respectively.

8. The acting surface of a tooth, projecting beyond the pitch circle, is called its face; that enclosed within the pitch circle, its flank.

9. The pitch of a wheel is the distance measured along the pitch circle from the face of one tooth to the corresponding face of the next; it includes, therefore, the breadth of a tooth and space. For two wheels to work in gear, the pitch must be the same in each.

10. Racks are toothed bars in which the pitch line is a straight line.

11. In annular wheels the teeth are cut on the internal edge of an annulus, or ring (fig. 92.)

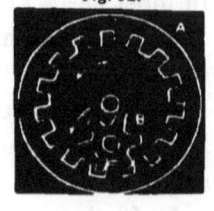

Fig. 92.

In fig. 93, B F is the line of centres; F A, A B, the primitive radii of the wheel and pinion respectively; A K L and A M N the pitch lines; K L and M N, the pitch; P L, the face; and Q L the flank, of the tooth.

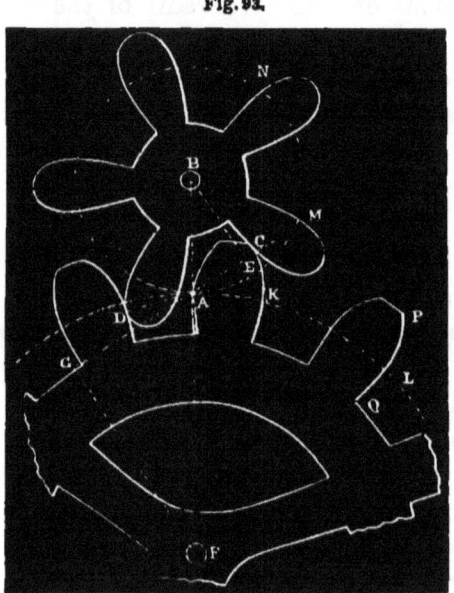

Fig. 93.

The pitch of Wheels.

We have seen that the pitch of a wheel is the

length of an arc of the pitch line comprising a tooth and space. Millwrights ordinarily measure the pitch as a cord of this arc, and, except in pinions with very few teeth, the two measurements sensibly coincide.

Having the diameter of a wheel, and the number of teeth, the pitch may be found, as follows:

Let D be the diameter of a wheel, N the number of teeth, and p the pitch; then, as $3{\cdot}1416$ D = the circumference of the circle,

$$p = \frac{3{\cdot}1416 \text{ D}}{\text{N}}$$

or approximately,

$$= \frac{22 \text{ D}}{7 \text{ N}}$$

Conversely, if the pitch of a wheel be given, and the number of teeth, then the diameter may be found,

$$\text{D} = \frac{p \text{ N}}{3{\cdot}1416} = \frac{7 \text{ N} p}{22} \text{ nearly.}$$

And if the pitch and diameter of a wheel be given, then the number of teeth may be found,

$$\text{N} = \frac{3{\cdot}1416 \text{ D}}{p} = \frac{22 \text{ D}}{7 p} \text{ nearly.}$$

But since a wheel must contain a whole number of teeth, N may never be a mixed number. If, therefore, this equation gives N with a fraction, a wheel cannot be constructed of that diameter and pitch. In this case, however, by slightly increasing or decreasing either the diameter or the pitch, the necessary conditions may be complied with.

In practice it is convenient to limit the number of pitches, with a view to the reduction of the number of patterns required for casting. Thus, the following series gives all the most ordinary pitches of my own practice:—

Spur flywheels, 5, 4½, 4, 3½, 3¼, 3, 2½, 2, 1½ inches.
Spur and bevel wheels, 5, 4½, 4, 3½, 3¼, 3, 2¾, 2½, 2¼, 2⅛, 2, 1¾, 1⅝, 1½, 1⅜, 1¼, 1⅛, 1, ⅞ inches.

Wheels of smaller pitch than this are not used in mill-work; but in machines, &c., the following pitches would probably be sufficient, viz:

1, ¾, ⅝, ½, ⅜, ¼, inch.

The value of $\pi = \frac{22}{7}$ ordinarily employed is not very accurate; hence it is convenient to calculate beforehand the values of $\dfrac{p}{3{\cdot}1416}$ and $\dfrac{3{\cdot}1416}{p}$ for the most useful pitches.

The following table gives these values:

Pitch in inches.	3·1416 / Pitch.	Pitch / 3·1416.	Pitch in inches.	3·1416 / Pitch.	Pitch / 3·1416.
5	0·6283	1·5915	1¾	1·7952	0·5570
4½	0·6981	1·4270	1⅝	1·9264	0·5141
4	0·7854	1·2732	1½	2·0944	0·4774
3½	0·8976	1·1141	1⅜	2·2848	0·4377
3¼	0·9666	1·0345	1¼	2·5132	0·3978
3	1·0472	0·9548	1⅛	2·7924	0·3580
2¾	1·1333	0·8754	1	3·1416	0·3182
2½	1·2566	0·7958	⅞	3·5904	0·2785
2¼	1·3963	0·7135	¾	4·1888	0·2396
2	1·5708	0·6366	⅝	5·0265	0·1988
1⅞	1·6755	0·5937	½	6·2832	0·1591

WHEELS AND PULLEYS.   119

RULE 1.—Given the pitch and number of teeth in a wheel to find its diameter.

Multiply the number of teeth by the constant in the third or sixth column of the preceding table corresponding to the pitch.

RULE 2.—Given the pitch and diameter of a wheel to find the number of teeth.

Multiply the diameter by the constant in the second or fifth column of the table corresponding to the pitch.

If this rule gives a mixed number, or whole number and fraction, a wheel cannot be constructed, as before said. The most convenient way of proceeding in that case will be to take the nearest whole number to the number given by the rule, and, using Rule 1, find a new diameter which will differ but slightly from the one previously assumed. This new diameter must be taken for the pitch circle in constructing the wheel.

Thus, suppose it required to find the diameter of a wheel of 2 inches pitch and 150 teeth. By Rule 1, we have $D = 150 \times 0.6366 = 95\frac{1}{2}$ inches $= 7$ ft. $11\frac{1}{2}$ inches.

Or, required the number of teeth in a wheel of 3 inches pitch and 9 feet diameter. By Rule 2 $N = 108 \times 1.0472 = 113.097$. Here the wheel will contain very nearly 113 teeth; but if we wish to know more accurately the diameter of a wheel of 3 inches pitch and 113 teeth, we find by the

1st Rule, $D = 113 \times 0.9548 = 107.89$ inches $= 8$ feet $11\frac{9}{10}$ inches. That is, a wheel of exactly 9 feet could not be constructed with a 3-inch pitch, but one of 8 feet $11\frac{9}{10}$ inches might and would contain 113 teeth.

Professor Willis has employed another method of graduating the sizes of wheels. Suppose the diameter, instead of the circumference, to be divided into as many equal parts as the wheel has teeth, and let one of these parts be called the diametral pitch of the wheel, to distinguish it from the common or circular pitch. Let M be the diametral pitch, so that

$$\frac{D}{N} = M$$

and let a series of values be taken for M in simple fractions of an inch, so that

$$M = \frac{1}{m}$$

where N and m are always whole numbers.

The ordinary values of m are 20, 16, 14, 12, 10, 9, 8, 7, 6, 5, 4, 3, 2, 1, which include wheels in which the circular or common pitch varies from $\frac{1}{8}$ inch to 3 inches, as shown in the following table, given by Professor Willis:

Value of m.	Circular Pitch in inches and decimals.	Circular Pitch to nearest one sixteenth	Value of m.	Circular Pitch in inches and decimals.	Circular Pitch to nearest one sixteenth
3	1·047	1	9	·349	—
4	·785	$\frac{3}{4}$	10	·314	$1\frac{5}{16}$
5	·628	$\frac{5}{8}$	12	·262	$\frac{1}{4}$
6	·524	$\frac{1}{2}$	14	·224	—
7	·449	$\frac{7}{16}$	16	·196	$1\frac{3}{16}$
8	·393	$\frac{3}{8}$	20	·157	$\frac{1}{8}$

This system is convenient where wheels of small pitch are employed, and involves less calculation than the common system.

Since $\frac{D}{N} = M$, we have $M = \frac{p}{3 \cdot 1416}$. Therefore, in the previous table (p. 118) the quantities in the third and sixth columns are the diametral pitches corresponding to the circular pitches in the first column, and the numbers in the second column are the corresponding values of m. In fact, this scheme differs from the first simply by expressing in small whole numbers the quantity $\frac{3 \cdot 1416}{p}$ instead of p.

The following table (pages 122 and 123) gives the relation of diameter, pitch, and number of teeth, for wheels of from $\frac{1}{2}$ inch to five inches pitch, and of from 12 to 200 teeth. Intermediate numbers may be found by direct proportion, by multiplying the number given for a wheel of half

TABLE SHOWING THE RELATION OF PITCH, DIAMETER, AND NUMBER OF TEETH.

No. of teeth in wheel.	Pitch in Inches.													
	½	¾	1	1¼	1½	1¾	2	2¼	2½	2¾	3	3¼	3½	4	4½	5
For each tooth add	·1591	·2386	·3182	·3978	·4774	·5570	·6366	·7135	·7958	·8754	·9548	1·035	1·114	1·273	1·427	1·592
12	1·91	2·86	3·82	4·77	5·73	6·68	7·64	8·56	9·55	10·50	11·46	12·41	13·37	15·28	17·12	19·10
13	2·07	3·10	4·14	5·17	6·21	7·24	8·28	9·28	10·35	11·38	12·41	13·45	14·48	16·55	18·55	20·69
14	2·23	3·34	4·46	5·57	6·68	7·80	8·91	9·99	11·14	12·26	13·37	14·48	15·60	17·83	19·98	22·28
15	2·39	3·58	4·77	5·97	7·16	8·36	9·55	10·70	11·94	13·13	14·32	15·52	16·71	19·10	21·41	23·87
16	2·55	3·82	5·09	6·37	7·64	8·91	10·19	11·41	12·73	14·01	15·28	16·55	17·83	20·37	22·83	25·46
17	2·70	4·06	5·41	6·77	8·12	9·47	10·82	12·13	13·53	14·88	16·23	17·59	18·94	21·64	24·26	27·06
18	2·86	4·30	5·73	7·17	8·60	10·03	11·46	12·84	14·32	15·76	17·19	18·62	20·05	22·92	25·69	28·65
19	3·02	4·54	6·05	7·56	9·08	10·58	12·10	13·56	15·12	16·63	18·14	19·66	21·17	24·19	27·11	30·24
20	3·18	4·77	6·36	7·96	9·55	11·14	12·73	14·27	15·92	17·51	19·10	20·69	22·28	25·46	28·54	31·83
21	3·34	5·01	6·68	8·36	10·02	11·70	13·37	14·98	16·71	18·38	20·05	21·72	23·40	26·74	29·97	33·42
22	3·50	5·25	7·00	8·76	10·50	12·25	14·01	15·70	17·51	19·26	21·01	22·76	24·51	28·01	31·39	35·01
23	3·66	5·49	7·32	9·16	10·98	12·81	14·64	16·41	18·30	20·13	21·96	23·79	25·62	29·28	32·82	36·60
24	3·82	5·73	7·64	9·55	11·46	13·37	15·28	17·12	19·10	21·01	22·92	24·83	26·74	30·56	34·25	38·20
25	3·97	5·97	7·96	9·96	11·94	13·93	15·92	17·84	19·90	21·89	23·87	25·86	27·85	31·84	35·68	39·79
30	4·77	7·16	9·55	11·93	14·32	16·71	19·10	21·41	23·87	26·26	28·64	31·04	33·42	38·21	42·81	47·74
35	5·57	8·35	11·14	13·92	16·71	19·50	22·28	24·97	27·85	30·64	33·42	36·21	38·99	44·57	49·95	55·70

WHEELS AND PULLEYS.

40	6·36	9·54	12·73	15·91	19·10	22·28	25·46	28·54	31·83	35·02	38·19	41·38	44·56	50·94	57·08	63·66
45	7·16	10·74	14·32	17·90	21·49	25·07	28·65	32·11	35·81	39·39	42·97	46·55	50·13	57·30	64·22	71·62
50	7·96	11·93	15·91	19·89	23·87	27·85	31·83	35·67	39·79	43·77	47·74	51·73	55·71	63·67	71·35	79·58
55	8·75	13·12	17·50	21·88	26·26	30·64	35·01	39·24	43·77	48·15	52·51	56·90	61·28	70·04	78·49	87·53
60	9·55	14·32	19·09	23·87	28·64	33·42	38·20	42·81	47·75	52·52	57·29	62·07	66·85	76·40	85·62	95·49
65	10·34	15·51	20·68	25·86	31·03	36·21	41·38	46·38	51·72	56·90	62·06	67·24	72·42	82·77	92·76	103·45
70	11·14	16·70	22·27	27·85	33·42	38·99	44·56	49·94	55·71	61·28	66·84	72·42	77·99	89·13	99·89	111·41
75	11·93	17·89	23·86	29·84	35·81	41·78	47·75	53·51	59·69	65·66	71·61	77·59	83·56	95·50	107·03	119·36
80	12·73	19·09	25·46	31·82	38·19	44·56	50·93	57·08	63·66	70·03	76·38	82·76	89·13	101·87	114·16	127·32
85	13·52	20·28	27·05	33·81	40·58	47·35	54·11	60·65	67·64	74·41	81·16	87·93	94·70	108·23	121·30	135·28
90	14·32	21·47	28·64	35·80	42·97	50·13	57·29	64·21	71·62	78·78	85·93	93·11	100·27	114·60	128·43	143·23
95	15·11	22·66	30·23	37·79	45·36	52·91	60·47	67·78	75·60	83·16	90·70	98·28	105·84	120·96	135·57	151·19
100	15·91	23·86	31·82	39·78	47·74	55·70	63·66	71·35	79·58	87·54	95·48	103·45	111·41	127·32	142·70	159·15
110	17·50	26·24	35·00	43·76	52·51	61·27	70·03	78·48	87·54	96·29	105·03	113·80	122·55	140·05	156·97	175·07
120	19·09	28·63	38·18	47·74	57·28	66·84	76·39	85·62	95·50	105·05	114·58	124·14	133·69	152·78	171·24	190·98
130	20·68	31·02	41·36	51·72	62·06	72·41	82·76	92·75	103·45	113·80	124·12	134·50	144·83	165·52	185·51	206·90
140	22·27	33·40	44·54	55·70	66·84	77·98	89·12	99·89	111·41	122·56	133·67	144·83	155·97	178·25	199·76	222·81
150	23·86	35·79	47·73	59·67	71·61	83·55	95·49	107·03	119·37	131·31	143·22	155·18	167·12	190·98	214·05	238·73
160	25·45	38·18	50·91	63·65	76·38	89·12	101·86	114·16	127·33	140·06	152·77	165·52	178·26	203·71	228·32	254·64
170	27·04	40·56	54·10	67·63	81·16	94·69	108·22	121·29	135·29	148·82	162·32	175·87	189·40	216·44	242·59	270·56
180	28·64	42·95	57·28	71·60	85·93	100·26	114·59	128·43	143·24	157·57	171·86	186·21	200·54	229·18	256·86	286·47
190	30·23	45·33	60·46	75·58	90·71	105·83	120·95	135·57	151·20	166·33	181·41	196·56	211·68	241·91	271·13	302·39
200	31·82	47·72	63·64	79·56	95·48	111·40	127·32	142·70	159·16	175·08	190·96	206·90	222·82	254·64	285·40	318·30

or a third of the number of teeth by two or three, or by adding together the diameters given for two wheels the sum of whose teeth is the number required. For an odd number of teeth, add the number given at the head of the table as many times as may be necessary to the diameter for a wheel of the nearest number of teeth given.

The Principles which Determine the Proper Form of the Teeth of Wheels.

The problem which presents itself in the construction of the teeth of wheels, is to discover the curvature which they should have in order that they shall revolve through the action of the teeth in precisely the same manner as they would by the rolling of the circumferences of their pitch lines.

The general principle by which this uniformity of motion is secured is as follows:—When wheels in gear act on each other so that a line perpendicular to the common tangent of the surfaces of the teeth at the point of contact passes always through the point where the pitch circles cut the line of centres, they will exert mutually the same force, move with uniform velocity, and be of true figure.

Or, in other words, the teeth will be rightly constructed when a line drawn from the point of contact of the pitch circles to the point of contact of two teeth is a normal to the surfaces in contact in all positions of the wheel and pinion.

Thus, let fig. 93 represent a wheel and pinion in gear, and let B A, A F be the primitive radii, and therefore A K L and A M N the pitch lines. Then if the teeth touch in C and D, and the lines A C, A D be always perpendicular to the common tangent to the touching parts, the teeth will be of true figure.

Epicycloidal Teeth

The epicycloid is the curve traced by a fixed point in the circumference of a circle, which rolls over or within the circumference of another circle, or on a straight line. Thus, let the circle A B C be fixed, and let the circle C D E roll over its circumference, then a point C in the circumference of this the generating circle will describe an epicycloid C, C', C'', C''', C'''', without the circle A B C. Similarly, a point F on the circumference of a generating circle F G, rolling within the circumference of A B C, will describe an interior epicycloid or hypocycloid F, F', F'', F'''.

The remarkable properties of the epicycloid which determine its fitness for describing the teeth of wheels are: 1st, when the generating circle is half the diameter of the base circle, and rolls within it, the hypocycloid is a straight line forming a diameter of the base; 2nd, if through the points of contact of the generating circle and the base, and the point describing the epicycloid, straight lines be drawn, these straight lines will be perpendicular to the curvature of the epicycloid from the point of contact B to the describing point at these points. Thus, for example, B C''' drawn

Fig. 94.

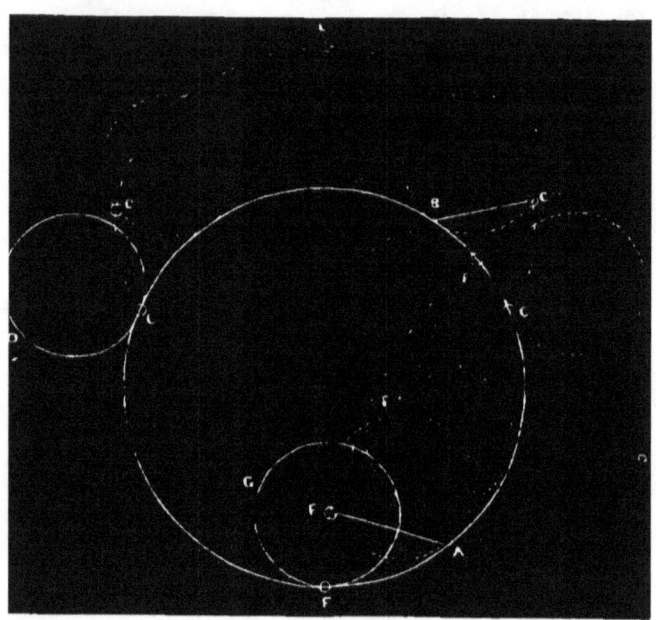

C''', is a normal to the curve at that point; and similarly A F' is a normal to the curve at F'.

Suppose in the same plane three circles R X Y (fig. 95), which touch each other in the point A, and whose centres F B G are consequently in a

WHEELS AND PULLEYS. 127

straight line. Let one of these circles be made to revolve round its centre, and force the other two to turn round their centres, which we suppose to

Fig. 95.

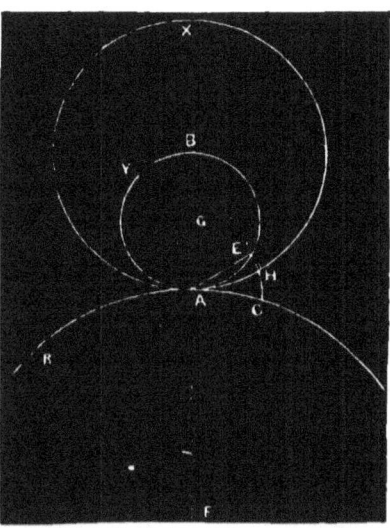

be fixed, moving these circles by the point of continual contact A, common to the three circumferences; it is evident that all the parts of the circumference of the circle made to revolve will be applied in succession to every part of the circumferences of the other two circles, in the same manner as if the two circles R and X remained immovable, while the third, Y, revolved on the circumferences of the other two. Hence, if we suppose a style fixed to the circumference of the circle Y, movable round its centre, the three circles having been obliged to turn by the motion of the one which has carried along the other two; when the style is at E, each of the two arcs A C and A H be made equal to the arc A E, the style will have described on the movable plane of the circle R, on the exterior part of which it revolves a portion C E of an epicycloid, and on the movable plane of the circle X, within which we may consider it to revolve, a portion E H of a hypocycloid. (*Camus*.)

These two epicycloids traced out at the same time by the style E affixed to the circle Y, will touch each other in the point E; for the straight line A E drawn through A, where the generating circle Y touches its bases R and X, will be a normal to the two epicycloids. The same will be true in every position of the circles, viz.: that the epicycloid and hypocycloid will have a common normal passing through A. Hence, if E C and E H be the faces of two teeth on the wheel and pinion R and X respectively, the condition of uniform motion already given will be complied with, the teeth will be of true form, and if the hypocycloid E H be moved by the epicycloid E C, or *vice versâ*, the wheel and pinion R and X will move precisely as if they rolled together at their pitch circles.

Wheels usually have their teeth constructed of such a form, that the flanks or parts within the pitch circle are bounded by straight lines radii of the pitch circles. Bearing in mind the property already stated, that the hypocycloid described by a generating circle of half the diameter of the base is a straight line forming a diameter of the base, we may so arrange our generating circle in describing the teeth of wheels as to comply with the above rule. By taking a generating circle Y of diameter equal to the radius of the base X, the hypocycloid E H will be part of a radius of X; or, in other words, a radius B H of X will always touch the epicycloid C E described without the circle R, by a generating circle Y, of a diameter equal to the radius of X. And the angle B E A being the angle of a semicircle, will always be a right angle. That is, the perpendicular to the straight line B H, at the point of contact with the epicycloid E C, will always pass through A.

We have hitherto supposed the circles moved by contact at the point A, in order to explain the generation of the epicycloid C E and straight line E H; but if we suppose these already described, the former being fixed to the circle R, and the latter to the circle X; then if E H roll by contact on the epicycloid C E, it will move the circle R precisely in the same manner as if the circle were moved by contact at A.

Construction of Epicycloidal Teeth.

Since every tooth in a wheel is of precisely the same form, it is sufficient to construct a single pattern tooth of true epicycloidal curvature, which may be used in setting out all the other teeth.

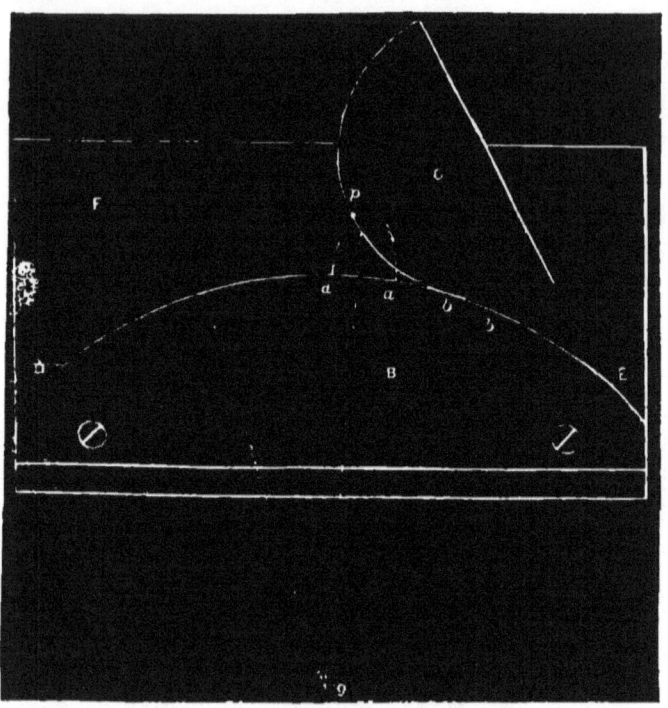

Fig. 96.

First method, when the generating circle is the same for wheel and pinion, the face of the tooth an epicycloid, and the flank a hypocycloid.

Construct two templets A and B (figs. 96, 97) having their faces arcs of the pitch circle of the wheel for which the tooth is required, and a third templet C cut to an arc of the intended generating

circle of the epicycloid. Fix a steel tracing point
p in the edge of the templet c, and for conveni-
ence a board F on which to draw the tooth, may
be fixed beneath the templet B. Mark off on the
board F (fig. 96) the pitch circle of the wheel D E,
and take distances $a\ b,\ b\ c$ equal to the pitch of
the teeth, and distances $a\ a',\ b\ b'$ equal to the
thickness of the teeth. If then the templet c be

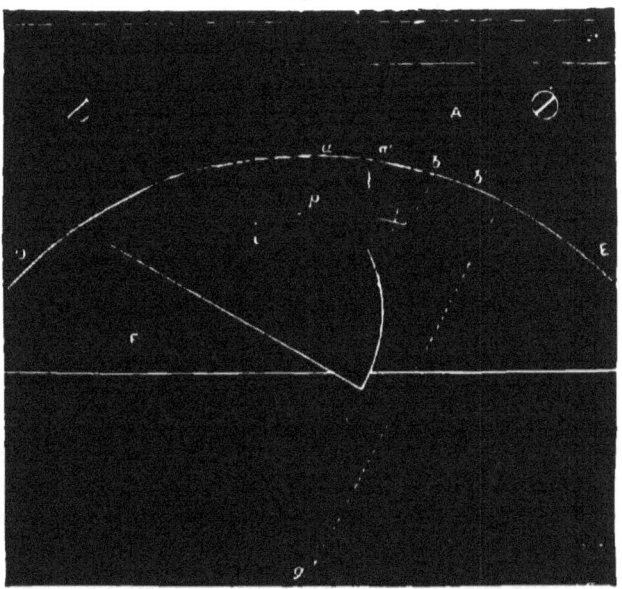

Fig. 97.

placed touching B, and with the tracing point p
coinciding with one of the marks as a, and be
rolled toward E, the point will trace out an epicy-

cloid $a\ p$ on the board F, which will form one face of the tooth. Next let the point p be made to coincide a', and the templet c be rolled toward D, the other face of the tooth will be described.

To draw the flanks, the templet A must now be fixed on the board F, with its face in contact with B; remove B and describe hypocycloids (fig. 97) from a and a', by rolling c on the inside of the pitch circle.

The length of the teeth is usually fixed as a proportional part of the pitch, but the least necessary length may be found experimentally by replacing the templet B on the board F, and making p coincide with a, roll c toward E till it touches B in b, the corresponding face of the next tooth; mark then the position of the tracing point and through this point draw an arc from the centre g of the wheel: this arc will mark the extremity of the tooth, and the arc $g\ p$ will be the true radius of the wheel.

This process, which, though complicated in description, is very easy in practice, must be repeated with two templets cut to the pitch circle of the pinion, the same generating circle c being employed; a similar pattern tooth will thus be found for the pinion, which will work with that already found for the wheel. The usual custom in practice is for the millwright first to describe the epicycloidal and hypocycloidal forms of the teeth required in the wheel and pinion; he then constructs two model teeth, one for the wheel and the other for the pinion, and from these he determines the true curves, and by means of his compasses transfers the same to the wheels or patterns on which these forms are to be impressed. The generating circle, it may be observed, must not exceed in size the radius of the pinion, or it would give rise to a weak form of tooth, thinner at the root than at the pitch circle.

Second method, where two generating circles are employed, in order that the flanks of the teeth may be straight lines radii of the wheel and pinion respectively.

It is the usual practice of millwrights to make the parts of the teeth of wheels within the pitch circles radii of the wheel. Now, we have seen that a hypocycloid described by a generating circle equal in diameter to the radius of the wheel would be a diameter of the wheel. If, therefore, the flank of the tooth of the wheel and the face of the tooth of the pinion be described by a templet cut to a radius equal to half that of the wheel and the flank of the tooth of the pinion and face of that of the wheel be described by a templet cut to a radius equal to half that of the pinion, then these teeth will work together truly, and will have radial flanks.

Since it is unnecessary to describe the flanks of such teeth by templets, there will be needed only one templet cut to the pitch circle of each wheel, but templets of two generating circles are required. In other respects the method is identical with that already described. The great defect of this method is, that neither the wheel nor pinion will work accurately with a wheel or pinion of any other diameter than that for which they were originally made, and thus a vast number of wheel patterns must be made to fulfil the requirements of practice; whereas wheels described by the previous method will work equally well with all other wheels the teeth of which have been described by the same generating circle—it being understood that only the parts of teeth *without* the pitch circle of the wheel roll on the parts *within* the pitch circle of the pinion, and those without the pitch circle of the pinion on those within the pitch circle of the wheel.

Hence Professor Willis has been led to suggest that for a given set of wheels a constant generating circle should be taken to describe both the parts without and within the pitch circles of the whole series, instead of making that circle depend on the diameters of the wheels. In this case the first solution must be employed, and the flanks of the teeth will not be straight; but the great advantage is gained, that any pair of wheels in the series will work together equally well.

To determine the proper size of the generating circle, we must remember that a tooth of weak form is produced when the generating circle is greater than half the diameter of the wheel. Hence the generating circle may be best made of a diameter equal to the radius of the smallest pinion of the series which are to work together.

The Rack is the extreme case of a wheel, or may be considered as a wheel of infinite radius. It may be described by either of the methods above, only noting that, if the second method be employed, the generating circle which traces the face of the teeth of the wheel becomes a straight line, and the epicycloid becomes an involute.

If the teeth of a series of wheels and of a rack be described by the same generating circle, any of the wheels will work with equal accuracy into the rack.

Involute Teeth.

The Involute.—The curve traced by a flexible line unwinding from the circumference of a circle, is called an involute.

Let P and W (fig. 98) be the pitch lines of a wheel and pinion, and let A and B be their centres. From A and B describe two circles D C, with radii A b and B b of the wheel and pinion respectively; so that

$$A c : B c :: A D : B C$$

Let $m\ n$ and $o\ p$ be two involute curves described by flexible lines unrolling from the circles D and C respectively, and touching at b. Then if b C, b D be drawn tangents to the circles at the points D and c, they are also in one straight line, because they are both normals to the curves at *b*. It may also be shown that the line C D intersects A B in *c*, where the pitch lines touch. Hence we have found two curves such, that the line perpendi-

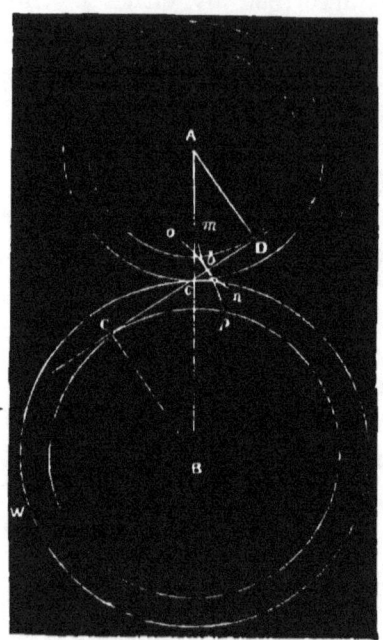

Fig. 98.

cular to their common tangent passes in all positions of the wheel and pinion through *c*, which is the sufficient condition of their uniform motion, if moved by the sliding of the curves instead of by contact at *c*. Hence, if the wheels be constructed with teeth formed to these involute curves, they will work with perfect regularity of motion.

In practice, the chief condition to be observed is to diminish the pressure on the axes, which is the chief defect of this form of teeth. The common tangent should be drawn through c, making an angle with A B, not deviating more than 20° from a right angle. Involute wheels have the double advantage that they work equally well if, through the wear of the brasses, the wheels have receded from one another; and any involute wheels of the same pitch and similarly described —that is, having the common tangent to the base circles passing through the point of contact of the pitch lines; or, in other words, base circles proportional to the primitive radii—will work together.

Mr. Hawkins, the translator of *Camus*, first proposed a simple instrument for describing the teeth of wheels to an involute curve. It consists of a straight piece of watch-spring $a\ b$ (fig. 99), with a screw at one end, and filed away at the edges so as to leave two teeth or tracers, $c\ c$, projecting from the edges of the watch-spring. At b a bit of wire is put through, and riveted, so as to form a knot by which the spring can be firmly held and stretched, as it is unwound from the base on which the involute is generated. This watch-spring is screwed to the edge of a templet A, curved to the radius of the base circle of the involute; and this being placed so that its centre coincides with the centre of the wheel, and revolved to bring one of the tracing points *c* in succession to each of the points at which corresponding faces of the teeth cut the pitch line, a

Fig. 99.

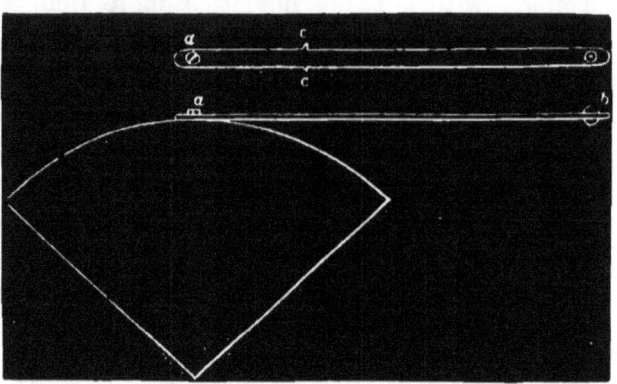

series of involute curves may be described by unfolding the watch-spring, whilst keeping it firmly stretched tangentially to the sector to which it is fixed. The sector A must then be turned over, and the involutes of the opposite faces of the teeth struck in a similar manner.

Another plan is to employ a straight ruler instead of the watch-spring, a tracer being fixed in its edge. This shows that the involute is an epicycloid generated by a straight line. The ruler must be kept in contact with the base circle, and the tracer brought in succession to all the points in which the faces of the teeth cut the pitch line.

Hence, to describe a wheel with involute teeth, the line of centres must be be drawn and divided proportionally to the number of teeth in the wheel and pinion. Draw the pitch line; divide the pitch line into the same number of equal parts as there are teeth in the wheel, and at these points mark out the thicknesses of the teeth all round. Draw the tangent to the base circles, making an angle of about 80° with the line of centres, which will give the radius of the base circle drawn touching it. A templet must be made to this radius, and then the involutes may be drawn by either of the preceding methods.

Allowance must be made to permit free play of the teeth in the spaces, the teeth being somewhat shorter than the distance between the bases of the involutes. But wheels of this figure require but little play in the engagement.

In the case of racks, the rack-teeth are bounded by straight lines perpendicular to the tangent drawn from the point where the pitch lines touch, to the base circle from which the involutes of the wheel are struck. If the teeth of the rack be made rectangular—that is, bounded by lines perpendicular to the pitch line—the involute must be struck from a base circle equal to the pitch circle of the wheel. In the former case there is a downward pressure on the rack; in the latter, the teeth of the wheel touch those of the rack in a single point—namely, the pitch line of the latter.

Professor Willis's Method of Striking the Teeth of Wheels.

In practice, the custom of describing the teeth of wheels as arcs of circles, has, from its simplicity, been generally adopted. The methods already given, however simple, when adopted in the formation of a single tooth, become tedious in their application to wheels of large size; and to this must be added the imperfect comprehension of their advantages by the millwrights charged with the task of designing wheel patterns.

Circular arcs struck at random, according to the judgment of the millwright, are often employed; and even where better principles have been introduced, it is common, after describing a single tooth accurately, to find by trial a circular arc nearly corresponding with its curve, and to employ this in marking out the cogs of the required wheel.

Seeing the advantages of the circular arc, and believing that it is not objectionable if only the employment of it is guided by true principles, Professor Willis has rendered this great service to practical mechanics—he has shown how, by a simple construction, the arcs of circles may be found, which, used in the construction of the teeth of wheels, will work truly on each other.

Let A B (fig. 100) be the centres of a wheel and pinion, and C the point of contact of the pitch circles on the line of centres. Through C draw c C c' at any angle with A B. Assume c as the centre from which to describe an arc for a tooth of the wheel a. Draw C D perpendicular to c C c', and from A through c draw A c D, meeting C D in

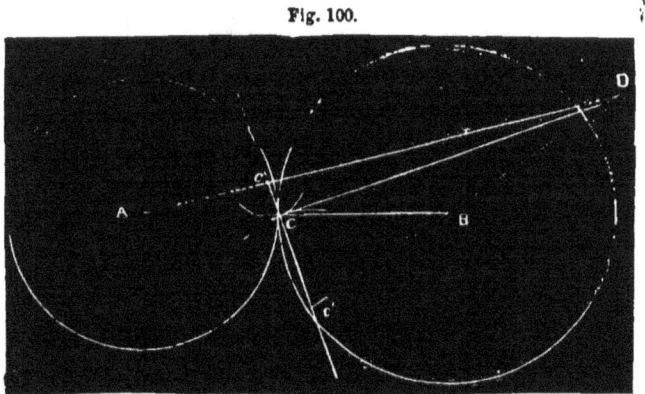

Fig. 100.

D. Lastly, from D through B draw D B c, meeting c C c' in c'. Then a small arc drawn from c with radius c C as a tooth for the wheel a, will work correctly with a small arc drawn from c', with a radius c' C as a tooth for the wheel B.*

Professor Willis recommends 75° 30' as the best magnitude of the angle A C c, so that Cos. 75° 30' = ¼. If this angle be constant in a set of wheels, any two will work truly together.

---

* Willis's "Principles of Mechanism," p. 123.

MACHINERY OF TRANSMISSION.

Fig. 101.

Tables showing the place of the Centres upon the Scales

Centres for the Flanks of Teeth

Number of Teeth	Pitch in inches							
	1	1¼	1½	1¾	2	2¼	2½	3
13	129	160	193	225	257	289	321	386
14	69	87	104	121	139	156	173	208
15	49	62	74	86	99	111	123	148
16	40	50	59	69	79	89	99	191
17	34	42	50	59	67	75	84	101
18	30	37	45	52	59	67	74	89
20	25	31	37	43	49	56	62	74
22	22	27	33	39	43	49	54	65
24	20	25	30	35	40	45	49	59
26	18	23	27	32	37	41	46	55
30	17	21	25	29	33	37	41	49
40	15	18	21	25	28	32	35	42
60	13	15	19	22	25	28	31	37
80	12	..	17	20	23	26	29	35
100	11	14	..	..	22	25	28	34
150	..	13	16	19	21	24	27	32
Rack	10	12	15	17	20	22	25	30

Centres for the Faces of Teeth

12	5	6	7	9	10	11	12	15
15	..	7	8	10	11	12	14	17
20	6	8	9	11	12	14	15	18
30	7	·9	10	12	14	16	18	21
40	8	..	11	13	15	17	19	23
60	..	10	12	14	16	18	20	25
80	9	11	13	15	17	19	21	26
100	..	..	..	..	18	20	22	..
150	..	..	14	16	19	21	23	27
Rack	10	12	15	17	20	22	25	30

The figure is of half the linear dimensions of the original.

Scale of Centres for Flanks of the Teeth

Scale of Centres for Faces of Teeth

For the easier description of these teeth, Professor Willis has invented the odontograph, a simple instrument of graduated card or wood, by which the position of the centres and radii of the arcs of the teeth can very easily be found. This instrument* is of the form shown in fig. 101, of half its proper lineal dimensions. It has the bottom edge bevelled off at an angle of 75°. The point where this would cut the right-hand edge is the zero of the scales. These scales are graduated to twentieths of an inch, to avoid fractional parts in the tables, and depart in each direction from the zero, the upper being that employed in finding the centres of the flanks of the teeth or parts within the pitch circle, and the lower for finding the centres of the faces of the teeth or parts without the pitch circle. Tables are given on the odontograph for finding the graduation on the scale corresponding to any given pitch and number of teeth. For intermediate pitches, not given in the table, or for wheels of greater size, the corresponding numbers can be found by simple proportion. For wheels of only twelve teeth the flanks are straight, and form parts of radii of the pitch circle.

In fig. 102, let A be the centre of a wheel, K d L the pitch line. Set off K L equal to the pitch, and

---

*Professor Willis's Odontograph may be obtained of Messrs. Holtzapfel of London.

bisect it in *d*. Draw radii A K, A L. Place the odontograph with its bevelled edge on the radius A K, and zero of the scale on the pitch line. Then look out in the table of centres for the flanks of teeth, the number corresponding to the pitch, and required number of teeth, and mark off this point *h*,

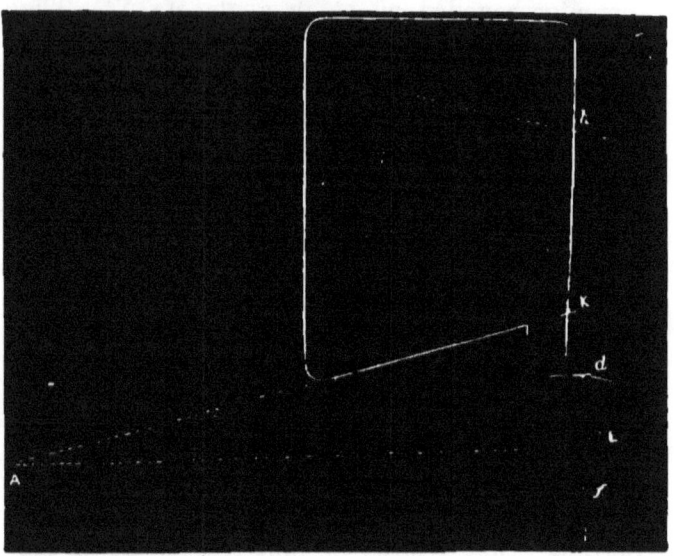

Fig. 102.

from the scale of centres for the flanks of teeth. Then remove the odontograph, and similarly place it on the radius A L. Find in the table of centres for the faces of the teeth the number corresponding to the pitch and number of teeth in the wheel, and mark it off at *f*, on the scale for centres of the faces of teeth. Then describe two arcs from *h* and f, with $h\,d$ and $f\,d$ as radii; these will form the side of a tooth. Then, from d let the pitch line be marked off into as many equal spaces as there are teeth in the wheel, and these be divided proportionally to the widths of the teeth and spaces. Through h and f, with radii $\text{A}\,h$ and $\text{A}\,f$, draw circles. Take $h\,d$ as a radius, and, placing one foot of the compass on the divisions of the pitch line, and the other in the circle drawn through h, describe a series of arcs forming the flanks of the teeth. Similarly with radius $f\,d$, and one leg of the compass on the circle drawn through f, describe the faces of the teeth.

For an annular wheel the same rules apply, only that the part of the curve which is face in a spur wheel becomes the flank in an annular wheel, and *vice versâ*. For a rack, the pitch line is straight, and A K, A L are parallel and perpendicular to it, at a distance equal to the pitch.

As these odontographs may be purchased in a very convenient form, with tables for their use, and also with tables of the widths of teeth, and spaces and length of teeth within and without the pitch circle, it is not necessary to describe them in further detail here.

General Form and Proportions of Teeth of Wheels.

The following have been drawn as a series of wheels and racks to illustrate the general form of the teeth of wheels. The pitch in figs. **103, 104,**

105, and 106 is one inch, and that in fig. 107 is 2½ inches.

In figs. 103, 104, 105, and 106 the wheel is 19·1 inches diameter; in fig. 107 it is 13 feet diameter.

Fig. 103 represents the form of the teeth on Professor Willis's system, the curves being arcs of circles. Fig. 104 gives the form of epicycloidal teeth, struck by a single generating circle rolled without the pitch circle for the faces, and within it for the flanks. This is the best system, as any pair of wheels so struck, with the same generating circle, and of equal pitch, will work together. Fig. 105 shows the common form of epicycloidal teeth, the flanks being straight. In this case the faces of the rack are struck by a generating circle half the diameter of the wheel, and the faces of the wheel, being obtained by a generating circle of infinite diameter or straight line, become involutes. Fig. 106 gives the form of teeth described as involutes, the curve being continuous, and, in the case of the rack, a straight line perpendicular to the tangent to the base circle. In these teeth it is possible to work with very little play. They are a good form for wheel and rack working together, the pressure on the journals being, in this case, less objectionable. Fig. 107 shows the teeth of a large wheel, traced from one of my own patterns, to exhibit the form and proportion which practice has shown to be desirable.

In these teeth the pitch $c\,d$ being 2½ inches, the

ON THE TEETH OF WHEELS. 147

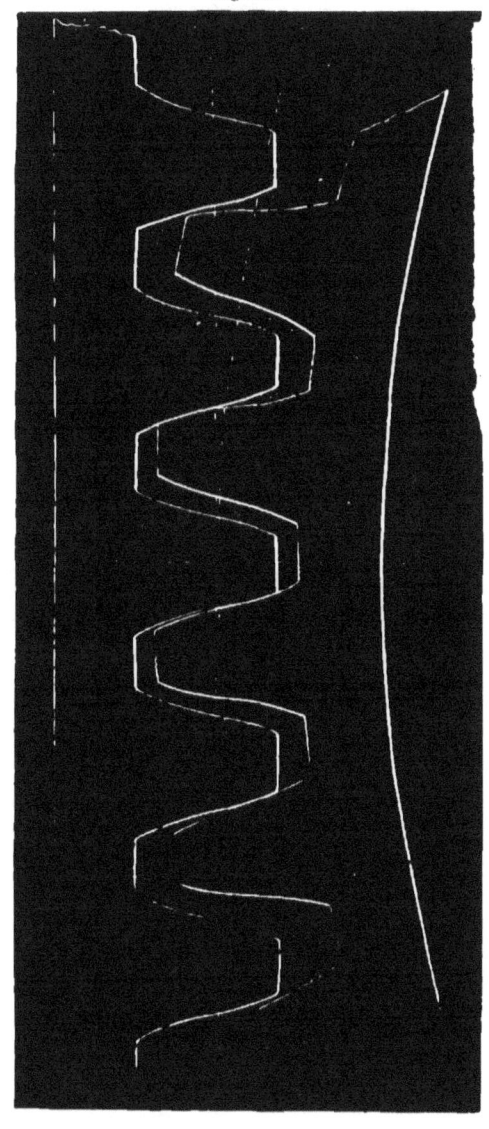

Fig. 101.

148 MACHINERY OF TRANSMISSION.

Fig. 104.

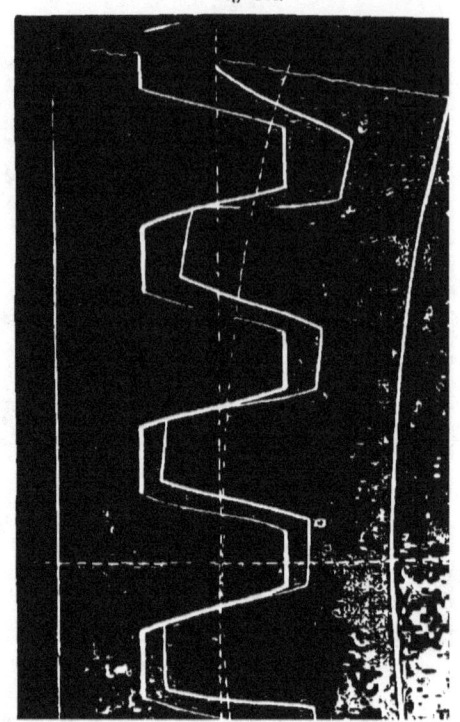

ON THE TEETH OF WHEELS. 149

Fig. 105.

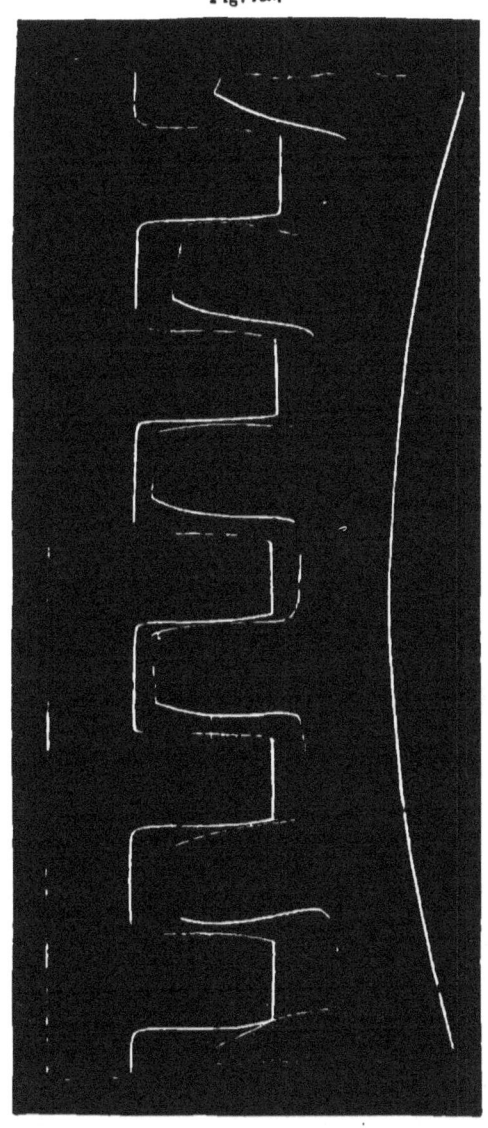

Fig. 106.

Fig. 107.

depth of the tooth or distance ab is $\frac{5}{8}$ths or $\frac{3}{4}$ths of the pitch. The proportions of the parts may be given as follows:—

			Proportional Part.	Inches.
Pitch	=	cd =	1·00	= 2½
Depth	=	ab =	0·75	= 1¾
Working depth	=	ae =	0·70	= 1⅝
Clearance	=	eb =	0·05	= ⅛
Thickness	=	cf =	0·45	= 1⅛
Width of space	=	fd =	0·55	= 1¼
Play or fd, cf	=		0·10	= ¼
Length beyond pitch line	=	ag =	0·35	= ¾

Taking these proportions we may construct a scale which shall give directly the corresponding numbers for any pitch. Taking a vertical line, and dividing it into eighths of an inch, we get the scale of pitches, (fig. 108.) Draw lines perpendicular to this, and on any one of them mark off a series of distances equal to the clearance, depth, thickness, etc., of the teeth corresponding to that pitch. Through o and these points draw the lines shown in the figure; they will divide the lines corresponding to all other pitches in the same proportion.

It is usual to allow a greater amount of clearance in small wheels than is necessary in large ones. Very varying proportions have been given by different millwrights, $\frac{1}{10}$th, $\frac{1}{12}$th, $\frac{1}{15}$th, and $\frac{1}{20}$th of the pitch having been used in different circumstances, even with the best mill-work. In the scale (fig. 108,) this has to a certain extent been taken into account; $\frac{1}{10}$th of the pitch is allowed

ON THE TEETH OF WHEELS. 153

Fig. 108.

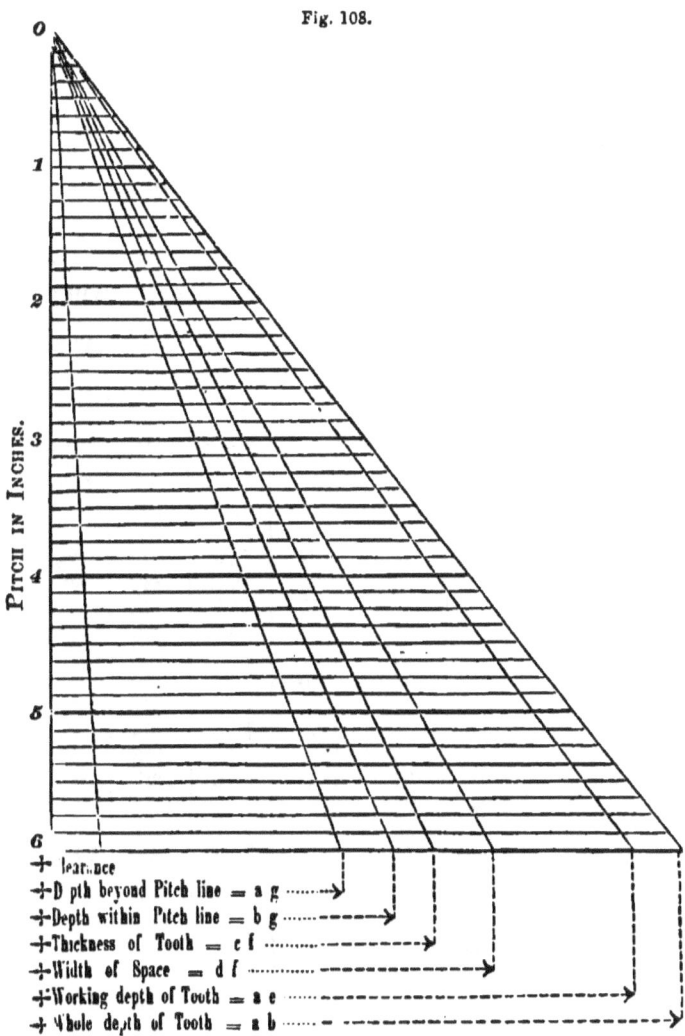

in smaller wheels, decreasing to $\frac{1}{15}$th in the largest; hence the lines are not absolutely straight, but are slightly curved, except that for the whole depth of the tooth, which quantity has been assumed to vary directly as the pitch.

Assuming that this scale represents with sufficient accuracy the proportions which practice shows to be best in average cases, we may construct a table for the guidance of the millwright. From this he must vary in cases where

TABLES OF PROPORTIONS OF TEETH OF WHEELS FOR AVERAGE PRACTICE.

Pitch.	Clearance and play.	Depth beyond pitch line.	Depth within pitch line.	Working depth.	Whole depth.	Thickness of tooth.	Width of space.
$\frac{1}{2}$	·06	·16	·22	·32	·38	·22	·28
$\frac{3}{4}$	·08	·25	·33	·50	·58	·33	·42
1	·10	·335	·435	·67	·77	·45	·55
$1\frac{1}{4}$	·12	·42	·54	·84	·96	·56	·69
$1\frac{1}{2}$	·13	·51	·64	1·02	1·15	·68	·82
$1\frac{3}{4}$	·14	·60	·74	1·20	1·34	·80	·95
2	·16	·685	·845	1·37	1·53	·92	1·08
$2\frac{1}{4}$	·17	·775	·945	1·55	1·72	1·04	1·21
$2\frac{1}{2}$	·19	·86	1·05	1·72	1·91	1·15	1·35
$2\frac{3}{4}$	·20	·95	1·15	1·90	2·10	1·27	1·47
3	·22	1·04	1·26	2·08	2·30	1·39	1·61
$3\frac{1}{4}$	·23	1·13	1·36	2·26	2·49	1·51	1·74
$3\frac{1}{2}$	·25	1·215	1·465	2·43	2·68	1·62	1·88
$3\frac{3}{4}$	·26	1·305	1·565	2·61	2·87	1·74	2·01
4	·28	1·39	1·67	2·78	3·06	1·86	2·14
$4\frac{1}{2}$	·31	1·565	1·875	3·13	3·44	2·09	2·40
5	·34	1·745	2·085	3·49	3·83	2·33	2·67
$5\frac{1}{2}$	·37	1·925	2·295	3·85	4·21	2·56	2·93
6	·40	2·10	2·50	4·20	4·60	2·80	3·20

ON THE TEETH OF WHEELS. 155

it appears necessary to allow more for defects of workmanship, or to permit less "backlash;"* it being understood that the table will only apply in cases where the teeth are formed with an approximation to the true mathematical figure.

In wood and iron gear where the teeth are carefully cut, very little if any clearance is necessary, as they work much better when the tooth of each wheel fills their allotted spaces. It is, however, different where wheels have to gear together direct from the foundry, where the teeth are not unfrequently deranged in the act of moulding in the sand.

This table gives the number to the nearest hundredth of an inch. It may be converted into the ordinary scale of eights by the following table:—

	Thirty Seconds of an Inch.									
	1	2	3	4	5	6	7	8	9	10
Corresponding Decimal.	·031	·062	0·94	·125	·156	·188	·219	·250	·281	·3125

As, unfortunately, decimal scales are not yet much used by millwrights, the following table has been prepared, giving the numbers in the preceding table in thirty seconds of an inch, such changes

---

* A technical expression for *reaction* on the back of the teeth.

being made as will reduce as much as possible the errors of employing this rough standard. The former table is to be preferred where it can be used, but in other cases the following one may be relied on. The left-hand figures in each column are inches, the right-hand ones thirty seconds of an inch, the denominators of the fraction being omitted.

TABLE GIVING THE PROPORTIONS OF TEETH OF WHEELS
IN INCHES AND THIRTY SECONDS OF AN INCH.

Pitch, inches.	Clearance.	Depth beyond the pitch line.	Depth within the pitch line.	Working depth.	Whole depth.	Thickness of tooth.
½	0″ 2	0″ 5	0″ 7	0″ 10	0″ 12	0″ 7
¾	0 3	0 8	0 11	0 16	0 19	0 10
1	0 3	0 11	0 14	0 22	0 25	0 14
1¼	0 4	0 13	0 17	0 26	0 30	0 18
1½	0 4	0 16	0 20	1 0	1 4	0 21
1¾	0 4	0 19	0 23	1 6	1 10	0 25
2	0 5	0 22	0 27	1 12	1 17	0 29
2¼	0 5	0 25	0 30	1 18	1 23	1 1
2½	0 5	0 28	0 33	1 24	1 29	1 5
2¾	0 6	0 31	0 37	1 30	2 4	1 8
3	0 7	1 1	1 8	2 2	2 9	1 12
3¼	0 7	1 4	1 11	2 8	2 15	1 16
3½	0 8	1 7	1 15	2 14	2 22	1 20
3¾	0 8	1 10	1 18	2 20	2 28	1 23
4	0 9	1 12	1 21	2 24	3 1	1 27
4½	0 10	1 18	1 28	3 4	3 14	2 3
5	0 11	1 24	1 35	3 16	3 27	2 10
5½	0 11	1 30	1 41	3 28	4 7	2 18
6	0 12	2 4	2 16	4 8	4 20	2 25

ON THE TEETH OF WHEELS. 157

Bevel Wheels.

Hitherto we have considered only that case of toothed wheels in which the pitch lines are in one plane. We have now to examine the modifications which are necessary when the axes of the

Fig. 109.

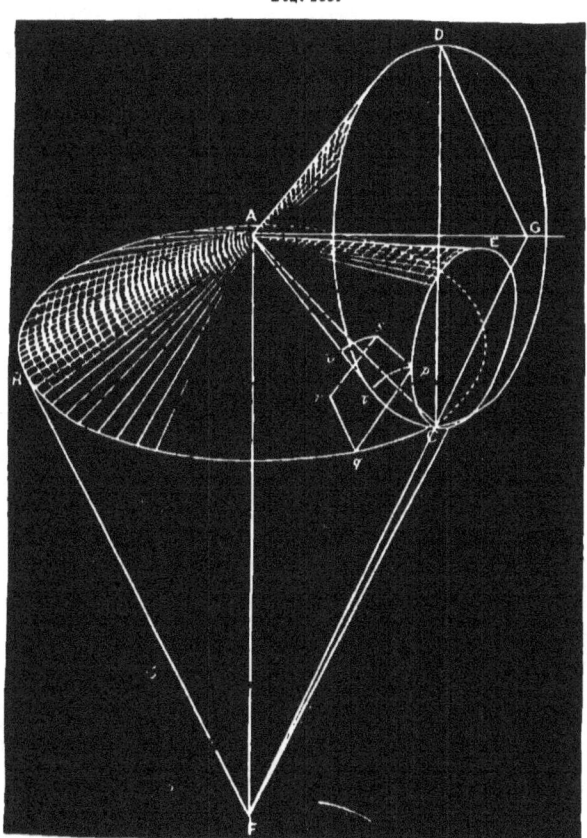

wheel and pinion arc inclined. It was shown in the preliminary chapter* that in this case motion might be transmitted by the rolling contact of the frustra of two cones. If, therefore, teeth be applied to these frustra, in the same manner as in spur gearing, they are attached to cylindrical surfaces, bevel gearing will be formed acting on the same principles of sliding contact which we have already discussed.

Let A B C, A C D (fig. 109) be two cones rolling in contact; take any other cone A E C also rolling in contact with A B C, in the line A C. As these cones roll together, the generating cone A E C will describe an epicycloidal surface *p q r s* on the outside of the cone A C D, and a hypocycloidal surface *p t v s* on the inside of the cone A C D. These surfaces will touch in the line *p s*, and will have a plane normal to their common tangent passing through A C. If, therefore, these surfaces be attached respectively to the cones A B C, A C D, and the motion of one cone be communicated to the other through the sliding contact of these surfaces, the motion will be uniform, as if the cones were driven by rolling contact at A C.

The curves *p t*, *p q*, lie in reality on the surface of a sphere of a radius equal to A C; but in practice, in bevel wheels, a small frustrum of a cone, tangential to the sphere at the circumference of

---

* See p. 66, § 88, 89.

the pitch line, is substituted for the spherical segment. Thus draw F C G (fig. 109) perpendicular to A C, cutting the axes of the cones in F and G. Let these lines revolve over the pitch lines of the cones and describe the narrow frustra. Then the epicycloidal surfaces may, without sensible error, be supposed to lie in these frustra, and to be generated there by the revolution of a generating circle C E.

Imagine the surface of these frustra to be unwrapped so as to lie in one plane, they will form parts of circular annuli. Thus let A B C, A C D (fig. 110), be two conical frustra; draw F C G as before, perpendicular to the line of contact A C. From G, with radii G H, G C, and G K, describe the circles K L, C M, H N; and from F, with radii F K, F C, F H, describe similar circles K P, C Q, H R; then the surfaces K P R H and K L N H will be developements of the frustra C D, C B. Let these be treated as spur wheels, and C Q, C M being treated as the pitch lines, let teeth be described by a describing circle in the method already explained for epicycloidal or other teeth. If, then, the plane on which these have been described, and which we suppose of drawing paper or other flexible material, be cut along the arcs K P, H R, K L, H N, the circular annuli may be wrapped round the frustra C B, C D, and the forms of the teeth traced off upon them.

The axes of bevel wheels are in practice, in

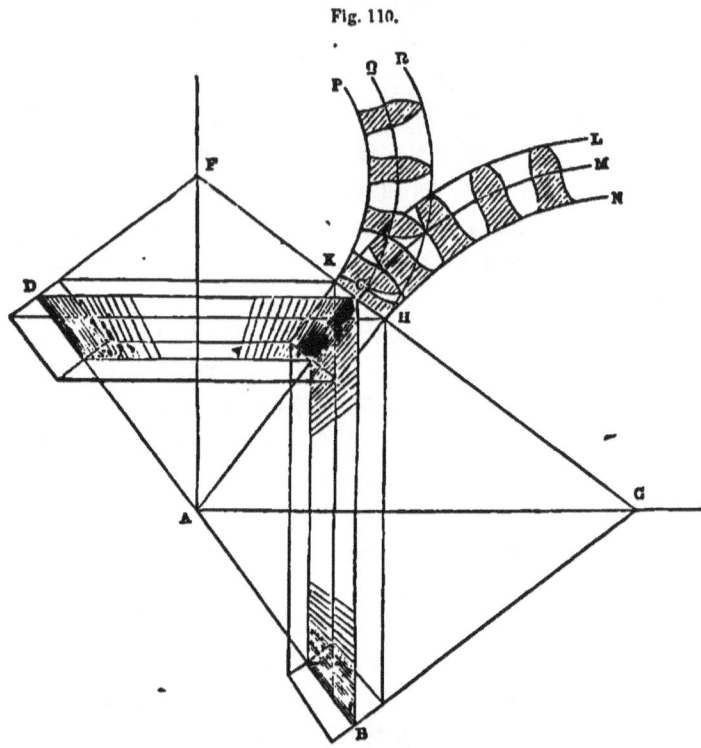

Fig. 110.

the great generality of cases, at right angles. Fig. 111 shows such a pair of bevels, with the frustra of the extremity of the teeth developed in the manner described.

Skew Bevels.

When two axes or shafts, which have to be connected by bevel wheels, do not meet in direc-

ON THE TEETH OF WHEELS. 161

tion, it is usual, as stated in the preliminary chapter,* to introduce an intermediate bevel wheel

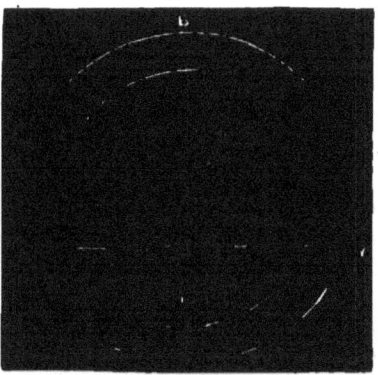

Fig. 111.

with two frustra. But the same object can more easily be accomplished by adopting skew bevels.

Let B p q (fig. 111) be the place of one of the two frustra, a its centre, and $a\,e$ the shortest distance between the axis of B p q, and the axis of the wheel to be connected with it. Divide $a\,e$ in c, so that $a\,c : e\,c ::$ mean radius of A B C : mean radius of frustrum working with A B C. Draw $c\,p\,q$ perpendicular to $a\,e$, then $c\,p$ or $c\,q$ is the line of action of the teeth, according to the direction in which the teeth are laid out in the pinion.

Figure 112 shows two wheels laid out in this manner; $a\,e$, as before, is the eccentricity or shortest distance between the two shafts, and is

---

* See page 68, § 70, 71.

14*

162   MACHINERY OF TRANSMISSION.

divided in *c* proportionally to the mean radii of the wheels; with centre *a* and radius *a c* describe a circle, and draw *e d* perpendicular to *a e*. Take

Fig. 112.

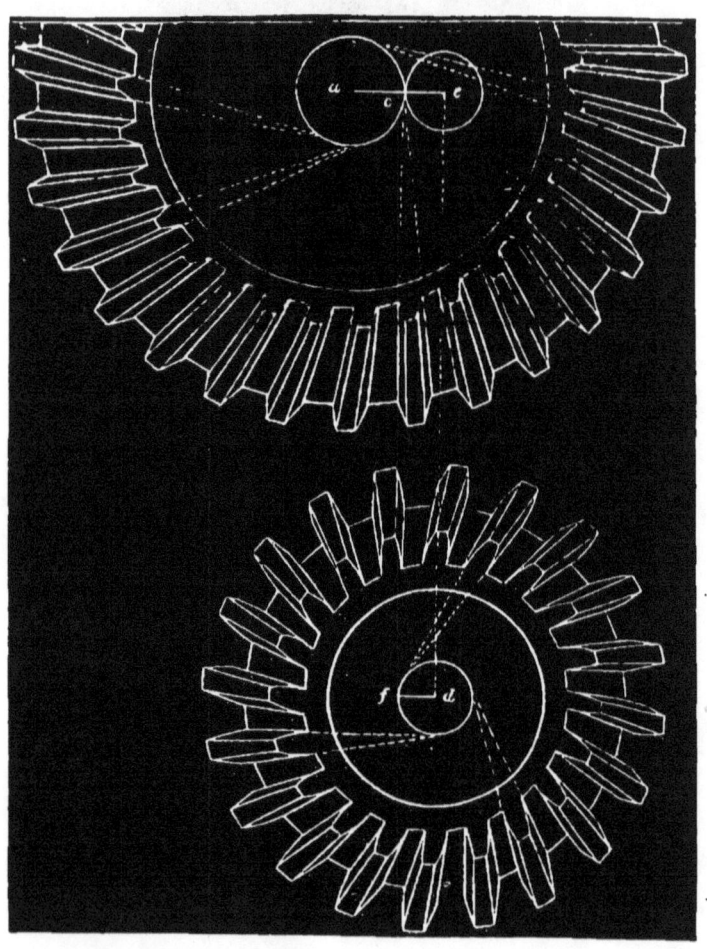

$df = ce$, then d will be the centre of the other wheel. From centre d, with radius fd, describe a circle. Then the directions of all the teeth in A B C will be tangents to the circle described about a, and the directions of all the teeth in D E F will be tangents to the circle described about f. Fig. 113 shows two such wheels in gear, the eccentricity permitting the shafts to pass each other.

Fig. 113.

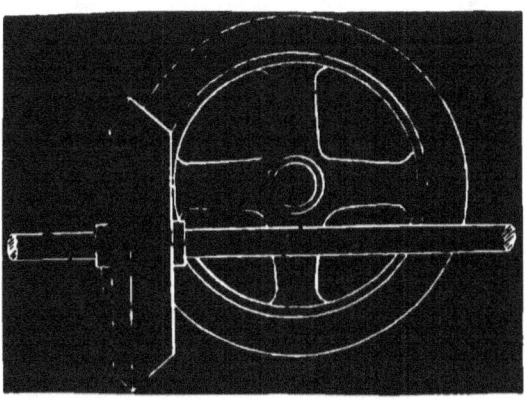

The Worm and Wheel.

By this contrivance the motion of a screw is communicated with great smoothness to oblique teeth on a spur wheel.

The section of a screw through its axis is precisely similar to that of a double rack. Let A B be such a section, and for simplicity suppose that the form of the threads of the screw has been determined by one of the rules already given for racks. Then the teeth of the wheel C D E may evidently be formed so as to work with the cen-

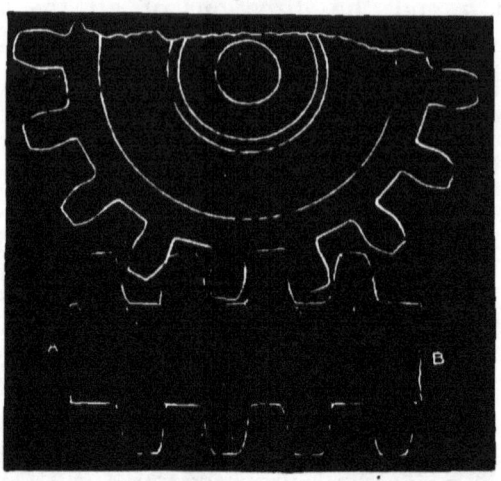

Fig. 114.

tre section of the screw. Now the effect of the revolution of the screw is precisely similar to that of the racks, and the sections of the threads of the screw will appear to travel from end to end, in the same way as a rack pushed forward in the same direction. If, therefore, it is sufficient that the wheel teeth be in contact with the screw at one point only, the teeth of the wheel may be made oblique, but straight, the obliquity being equal to the pitch of the screw. This is the usual practice of millwrights. If, however, the teeth are required to be in contact with the entire breadth of the tooth, the outline of the tooth must vary in every section of the wheel, and the process of describing these teeth becomes very complex. Practically, the difficulty has been overcome by first making a pattern screw of steel, notched in the threads to convert it into a cutting instrument. The wheel is then roughly cut out, and being fixed in a frame, the screw is used to cut out the spaces between the teeth to their true form.

Strength of the Teeth of Wheels.

The pressure on the teeth varies directly as the horse-power transmitted and inversely as the velocity of revolution. Thus if one wheel transmit 5 horse-power and another 10 horse-power at the same velocity, the strain on the latter will be twice that on the former. Or, again, if two wheels transmit the same power, but one at a velocity of 100 feet per minute, and the other at only 25 feet per minute, the strain on the former will be only one-fourth that on the latter.

Let v be the velocity in feet per second, H the number of horse-power transmitted, then the total pressure on the wheels will be—

$$P = \frac{550\,H}{v}$$

where P is the statical pressure in lbs.

For example, suppose the fly-wheel of an engine to be 24 feet in diameter, and to work into a pinion 5 feet diameter. And let the work transmitted be 150 horse-power. Then, if the wheel makes 25 revolutions per minute, the periphery will move at a velocity of $\dfrac{75\cdot 4 \times 25}{60} = 31\cdot 4$ feet per second; and the statical pressure on the teeth will be $\dfrac{550 \times 150}{31\cdot 4} = 2627$ lbs.

In addition to statical pressure, however, a different element has to be taken into account, namely, the impacts due to sudden accelerations or retardations of speed. The allowance which must be made to prevent accident from this cause varies exceedingly in different kinds of machinery. It is great in the gearing of rolling mills for instance, and in all machinery in which the strains are irregular.

In calculating the strength of the tooth, it has been usual to consider it as a short beam fixed at one end, and having the whole of the pressure applied along the extremity of the tooth. But there is a position in which the teeth may be subjected to a severer stress still; owing to the wear of brasses

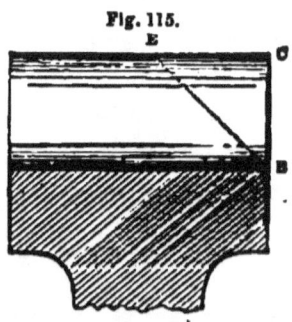

Fig. 115.

and teeth, we cannot calculate upon the strain

bearing always on the whole breadth of the tooth. The pressure may not only come on to the extremity of a tooth, but if any obstruction come in between the teeth, it may be thrown entirely upon one corner of the tooth. In such a case it may be shown, by the rules of maxima and minima, that if $EC = CB$, the greatest stress will be near the line EB.

Tredgold has expressed the strength of a tooth on this supposition by the formula

$$W = \frac{f d^2}{5}$$

where d is the thickness of the tooth. To allow for wear, however, he adds one-third, so that

$$W = \frac{f d^2 (1 - \frac{1}{3})^2}{5} = \frac{f d^2}{10 \cdot 25}$$

In cast-iron $f = 15,300$, and hence

$$d = \sqrt[2]{\frac{W}{1500}}$$

Or in words, the thickness necessary for the tooth or inches is equal to the square root of the stress on the tooth in pounds divided by 1500. Hence Tredgold has computed the following table, the breadths of the teeth being deduced, on the principle that the stress should not exceed 400 lbs. per inch breadth:—

TABLE OF THICKNESS, BREADTH, AND PITCH OF TEETH OF WHEELS.

Stress in lbs. at the pitch line.	Thickness of teeth in inches.	Breadth of teeth in inches.	Pitch in inches.
400	0·52	1	1·1
800	0·73	2	1·5
1,200	0·90	3	1·9
1,600	1·03	4	2·2
2,000	1·15	5	2·4
2,400	1·26	6	2·7
2,800	1·36	7	2·9
3,200	1·46	8	3·0
3,600	1·56	9	3·3
4,000	1·64	10	3·4
4,400	1·70	11	3·6
4,800	1·78	12	3·7
5,200	1·86	13	3·9
5,600	1·93	14	4·0
6,000	2·00	15	4·2

To use this table when the horses' power transmitted by the wheel are known, the reader must refer to the table on page 172.

Elsewhere Tredgold has given a rule of the following description:—

$$d = \tfrac{3}{4} \sqrt{\frac{H}{v}} \text{ for cast-iron,}$$

where d is the requisite thickness of a tooth to transmit a force of H horses at a velocity v feet per second.

Hence Tredgold's last rule for the thickness of cast-iron teeth is as follows —" Find the number of horses' power transmitted by the wheel, and divide that number by the velocity in feet per second of the pitch line of the pinion or wheel; extract the square root of the quotient, and three fourths of this root will be the least thickness of cast-iron teeth for the wheel or pinion." From this he derives a second rule for the pitch, which manifestly depends on the thickness of the tooth, namely, multiply the thickness of the tooth by 2·1 and the product will be the pitch. The same result may be obtained from inspection of the tables I have given at pages 154, 156. Wooden teeth he recommends to be made of twice the thickness of cast-iron ones. But one-and-a-half times the thickness is a sufficient allowance.

A writer in the "Engineer and Machinists' Assistant" deduces another but equally simple rule for the thickness of teeth; he assumes the relation

$$t = c\sqrt{w};$$

where t is the thickness of the tooth, w the pressure on the tooth and c a constant, depending on the nature of the material. Let then a be the strength of a bar 1 inch long, 1 broad, and 1 thick. Then, to support a weight w by a bar of a length l, and breadth b,

$$t = \sqrt{\frac{w \times l}{a \times b}};$$

suppose the breadth of the tooth to be fixed at twice its length;

$$t = \sqrt{\frac{w\,l}{a \times 2l}} = \sqrt{\frac{w}{2a}}.$$

Taking $a = 8000$ lbs. for cast-iron, $2a = 16,000$ lbs., but as this is the breaking weight, the safe working-pressure will be only 1600 lbs., and the thickness of the tooth for safe working will be for cast-iron:

$$t = \sqrt{\frac{w}{1600}} = 0{\cdot}025\,\sqrt{w}.$$

Where w being given in lbs. t is found in inches. Similarly for other materials he obtains:

$$c = {\cdot}035 \text{ for brass,}$$
$$= {\cdot}038 \text{ for hard wood.}$$

For example, in the wheel assumed at p. 164, w was found to be 2627 lbs. Hence the necessary thickness of the tooth, if of cast-iron, would be ·025 $\sqrt{2627}$ = 1·28 inches. Referring to the tables of the relation of pitch, etc., we find that the wheel must be of 2¾ inches pitch, the teeth of 2·1 inches length, and the breadth of the wheel 2·1 × 2 = 4·2 inches at the least. By Tredgold's latter rule, the thickness of the teeth for the same wheel would be $t = \frac{3}{4}\sqrt{\frac{150}{31{\cdot}4}}$ = 1·41 inches: the pitch = 2·1 × 1·41 = 3·0 inches, and the breadth = $\frac{2627}{400}$ = 6½ inches.

Bearing in mind that $w = \dfrac{550 \text{ H}}{v}$, where H is the maximum horse-power transmitted, and v the velocity of the pitch line of the wheel in feet per second, we may give these formulæ in a more convenient form:

$$t = x \sqrt{\dfrac{H}{v}}.$$

Where $x = 0\cdot587$ for cast-iron,
" $= 0\cdot821$ for brass,
" $= 0\cdot891$ for wood.

Conversely, if a wheel having teeth t inches thick be given, the horse power it is capable of transmitting is given by the formula:

$$H = \dfrac{t^2 v}{x^2}.$$

Where $x^2 = 0\cdot344$ for cast-iron,
" $= 0\cdot674$ for brass,
" $= 0\cdot795$ for wood.

From the following table the pressure at other velocities, and with another amount of horse-power, may be obtained by interpolation, remembering that the pressure varies inversely as the former, and directly as the latter. To this we have appended another table, giving the horses' power, which can be safely transmitted by wheels of different pitches when proportioned according to the table at page 154. The last of these tables has been calculated on the assumption that 400 lbs. per inch is the greatest working stress which is consistent with durability in ordinary cases.

RELATION OF HORSES' POWER TRANSMITTED AND VELOCITY AT THE PITCH CIRCLE TO PRESSURE ON TEETH.

Number of horses' power transmitted.	\multicolumn{10}{c|}{Velocity in feet per second.}									
	1 ft.	3 ft.	5 ft.	7 ft.	9 ft.	11 ft.	13 ft.	15 ft.	20 ft.	25 ft.
lbs.	lbs.	lbs.	lbs.	lbs.	lbs.	lbs.	lbs.	lbs.	lbs.	lbs.
1	550	183	110	79	61	50	42	37	28	22
2	1,100	367	220	157	122	100	85	73	55	44
3	1,650	550	330	236	183	150	127	110	83	66
4	2,200	733	440	314	244	200	169	146	110	88
5	2,750	917	550	393	306	250	212	183	138	110
6	3,300	1,100	660	471	367	300	254	220	165	132
10	5,500	1,833	1,100	786	611	500	423	367	275	220
15	8,250	2,750	1,650	1,179	917	750	635	550	413	330
20	11,000	3,667	2,200	1,571	1,222	1,000	846	733	550	440
25	13,750	4,583	2,750	1,964	1,527	1,250	1,058	917	688	550
30	16,500	5,500	3,300	2,357	1,833	1,500	1,269	1,100	825	660
40	22,000	7,333	4,400	3,143	2,444	2,000	1,692	1,467	1,100	880
50	27,500	9,167	5,500	3,928	3,055	2,500	2,115	1,833	1,375	1,100
60	33,000	11,000	6,600	4,714	3,667	3,000	2,538	2,200	1,650	1,320
70	38,500	12,833	7,700	5,500	4,278	3,500	2,962	2,567	1,925	1,640
80	44,000	14,667	8,800	6,285	4,889	4,000	3,385	2,933	2,200	1,760
90	49,500	16,500	9,900	7,071	5,500	4,500	3,808	3,308	2,475	1,980
100	55,000	18,333	11,000	7,857	6,111	5,000	4,231	3,667	2,750	2,200
110	60,500	20,167	12,100	8,643	6,722	5,500	4,654	4,033	3,025	2,420
120	66,000	22,000	13,200	9,423	7,333	6,000	5,077	4,400	3,300	2,640
130	71,500	23,833	14,300	10,214	7,944	6,500	5,500	4,767	3,575	2,860
140	—	25,667	15,400	11,000	8,556	7,000	5,923	5,133	3,850	3,080
150	—	27,500	16,500	11,786	9,167	7,500	6,346	5,500	4,125	3,300
160	—	29,333	17,600	12,571	9,778	8,000	6,769	5,867	4,400	3,520
170	—	31,167	18,700	13,357	10,389	8,500	7,192	6,233	4,675	3,740
180	—	—	19,800	14,143	11,000	9,000	7,615	6,600	4,950	3,960
190	—	—	20,900	14,929	11,611	9,500	8,038	6,967	5,225	4,180
200	—	—	22,000	15,714	12,222	10,000	8,462	7,333	5,500	4,400
300	—	—	33,000	23,571	18,333	15,000	12,692	7,700	8,250	6,600
400	—	—	44,000	31,428	24,444	20,000	16,923	8,067	11,000	8,800
500	—	—	55,000	39,285	30,555	25,000	21,154	8,433	13,750	11,000

ON THE TEETH OF WHEELS.

TABLE SHOWING THE PITCH AND THICKNESS OF TEETH TO TRANSMIT A GIVEN NUMBER OF HORSES' POWER AT DIFFERENT VELOCITIES.

Pitch in inches.	Thickness of teeth in inches.	\multicolumn{12}{c}{Velocity at Pitch line in feet per second.}
		1	3	5	7	9	11	13	16	20	25	30	40	60
		H.P.	H.P.	H.P.	H.P.	H.P.	H.P.	H.P.	H.P.	H.P.	H.P.	H.P.	H.P.	H.P.
½	·22	·14	·42	·71	·99	1·26	1·5	1·8	2·1	2·8	3·5	4·2	5·6	7·1
¾	·33	·32	·95	1·59	2·22	2·75	3·5	4·1	4·7	6·3	7·9	9·5	12·7	16·9
1	·45	·59	1·77	2·95	4·12	5·31	6·5	7·7	8·8	11·8	14·7	17·7	23·6	29·6
1¼	·56	·91	2·74	4·56	6·38	8·22	10·0	11·9	13·7	18·2	22·8	27·4	36·5	45·6
1½	·68	1·34	4·03	6·72	9·41	12·09	14·8	17·5	20·2	26·9	33·6	40·3	53·7	67·2
1¾	·80	1·86	5·58	9·31	13·03	16·74	20·5	24·2	27·9	37·2	46·6	55·8	74·4	93·1
2	·92	2·46	7·38	12·31	17·23	22·14	27·1	32·0	36·9	49·2	61·5	73·8	98·4	123·1
2¼	1·04	3·14	9·43	15·72	22·01	27·29	34·6	40·9	47·2	62·9	78·6	94·3	125·7	157·2
2½	1·15	3·85	11·54	19·23	26·92	34·62	42·3	50·0	57·7	76·9	96·1	115·4	153·8	192·3
2¾	1·27	4·69	14·07	23·46	32·82	42·21	51·6	61·0	70·3	93·8	117·2	140·6	187·5	234·5
3	1·39	5·62	16·84	28·08	39·31	50·42	61·8	73·0	84·2	112·3	140·4	168·5	224·6	280·8
3¼	1·51	6·63	19·88	33·14	46·40	59·64	72·9	86·2	99·4	132·6	165·7	198·8	265·1	331·4
3½	1·62	7·63	22·89	38·15	53·40	68·67	83·9	99·2	114·4	152·6	190·7	228·9	305·1	381·5
3¾	1·74	8·80	26·40	44·00	61·61	79·20	96·8	114·4	132·0	176·0	220·0	264·0	352·0	440·0
4	1·86	10·06	30·17	50·29	70·40	90·50	110·6	130·7	150·8	201·1	251·4	301·7	402·3	502·9
4½	2·09	12·70	38·09	63·49	88·89	114·27	139·7	165·1	190·5	254·0	317·5	380·9	507·9	634·9
5	2·33	15·78	47·35	78·91	110·47	142·05	173·6	205·2	236·7	315·6	394·5	473·5	631·3	789·1
5½	2·56	19·05	57·15	95·26	133·36	171·45	209·6	267·7	285·8	381·0	476·3	571·5	762·0	952·6
6	2·80	22·79	68·37	113·96	159·54	205·11	250·7	316·3	341·9	455·8	569·7	683·7	911·6	1,139·6

TABLE SHOWING THE BREADTH OF TEETH REQUIRED TO TRANSMIT DIFFERENT AMOUNTS OF FORCE AT A UNIFORM PRESSURE OF 400 LBS. PER INCH.

Number of horses' power transmitted.	Velocity of pitch line in Feet per second.											
	1	3	5	7	9	11	13	15	20	25	30
	ins.	ins.	ins.	ins.	ins.	ins.	ins.	ins.	ins.	ins.	ins.
1	1·4	0·5	0·3	0·2	0·15	0·12	0·11	0·09	0·07	0·05	0·05
2	2·7	0·9	0·6	0·4	0·3	0·25	0·21	0·18	0·14	0·11	0·09
3	4·1	1·4	0·8	0·6	0·5	0·38	0·32	0·27	0·21	0·17	0·14
4	5·5	1·8	1·1	0·8	0·6	0·50	0·42	0·37	0·28	0·22	0·18
5	6·9	2·3	1·4	1·0	0·8	0·62	0·53	0·46	0·35	0·28	0·23
10	13·7	4·6	2·7	2·0	1·5	1·3	1·06	0·92	0·69	0·55	0·46
15	20·6	6·9	4·1	2·9	2·3	1·9	1·6	1·37	1·03	0·83	0·69
20	34·2	9·2	5·5	3·9	3·1	2·6	2·1	1·83	1·38	1·10	0·92
25	41·2	11·4	6·9	4·9	3·8	3·1	2·6	2·3	1·72	1·38	1·14
30	55·0	13·8	8·2	5·9	4·6	3·8	3·2	2·8	2·06	1·65	1·38
40	68·8	18·3	11·0	7·9	6·1	5·0	4·2	3·7	2·75	2·20	1·83
50	82·5	22·0	13·7	9·8	7·6	6·3	5·3	4·6	3·44	2·75	2·29
60	96·2	27·5	16·5	11·8	9·1	7·6	6·4	5·5	4·1	3·30	2·75
70	110·0	32·1	19·2	13·8	10·7	8·8	7·4	6·4	4·8	3·85	3·21
80	123·8	36·7	22·0	15·7	12·2	10·0	8·5	7·3	5·5	4·40	3·67
90	137·5	41·2	24·8	17·7	13·8	11·2	9·6	8·2	6·2	4·95	4·12
100	151·2	45·8	27·5	19·6	15·3	12·5	10·6	9·1	6·9	5·50	4·58
110	165·0	50·4	30·3	21·6	16·8	13·7	11·6	10·1	7·5	6·05	5·04
120	178·8	55·0	33·0	23·5	18·3	15·0	12·7	11·0	8·3	6·6	5·50
130	—	59·8	35·7	25·5	19·8	16·3	13·8	11·9	8·9	7·2	5·98
140	—	64·2	38·5	27·5	21·4	17·5	14·8	12·8	9·6	7·7	6·42
150	—	68·7	41·3	29·4	22·9	18·7	15·8	13·7	10·3	8·2	6·87
160	—	73·3	44·0	31·4	24·5	20·0	16·9	14·6	11·0	8·8	7·33

CHAPTER III.

ON THE STRENGTH AND PROPORTIONS OF SHAFTS.

THE system of transmitting power from a common centre to a large number of machines, at some distance, is comparatively modern. In the operations of spinning and weaving by a consecutive series of machines, placed in rows, shafting became essential for distributing the power of the common prime mover. At first, the machines were brought as close to the prime mover as possible; and the early construction of mills—when the water-power was divided into separate falls—must be fresh in the recollection of many persons now living. In some cases, before the introduction of the steam engine, it was the custom to have a separate water-wheel to every machine, thus splitting up the power into as many parts as there were machines, or pairs of machines, to drive. In process of time, it was found more convenient, on the score of economy, to husband the water and concentrate the prime movers; hence one large water-wheel was constructed, around which the machinery was arranged, either in rows or otherwise as best suited the work to be performed.

This principle, of the concentration of the motive power, destroyed the old system of separate buildings, and led to the employment of a large number of machines for the various processes of manufacture in one building. From this we derive the Factory system, in which any number of processes are carried on, the machinery being distributed over the different floors of a large building, and receiving motion from a single prime mover at a convenient distance. In this way, the power is conveyed by lines of shafting coupled together in lengths, adapted to the bays or divisions of the building. At first, the buildings were short, and shafting of great length was not required; gradually, more and more machines were concentrated in the same building, and shafting of 200 or 300 feet in length became necessary. To show to what an extent this system has been carried, it may be mentioned that, in the large mills at Saltaire, the shafting, if placed in a single line, would extend for a distance of more than two miles. This progress has been chiefly due to the introduction of the steam engine, in place of waterwheels, because the available power is no longer limited by the circumstances of the locality in which the mill is placed.

This concentration of a great number of machines in one building is peculiar to the Factory system; and in the present highly-improved state of mechanical science and its application to the production of textile fabrics, it has become essential to economy in the manufacturing processes, that they should be carried on in the same building. Spinners and manufacturers are fully aware of the advantages peculiar to this system of concentration, so much so, that out of what would formerly have been considered a mere fractional saving, large profits and large fortunes are now made. In fact, the amalgamation of the different processes under one management and under one roof, gave rise to the shed system, where the operations of the manufacture of cotton are carried on under what is called the "*sawtooth*" roof, in order to bring the whole on the ground-floor under one inspection.

1. *The Material of which Shafting is constructed.*

The selection of the material for shafting is of great importance, and the uses to which it is to be applied require careful consideration. Formerly wood, with iron hoops and gudgeons, was universally employed; then cast-iron was introduced and subsequently wrought-iron has in most cases superseded both. Wood, indeed, has become obsolete; but cast-iron is as good as, if not superior to, wrought-iron, in certain cases. The main and vertical shafts of a mill are generally of cast-iron, both on account of its cheapness, and its high resistance to torsion. The vertical shafts, which convey the power from the first motion wheels to the different rooms of the mill, are more rigid and less subject to vibration when of cast-iron;

even the main horizontal shafting, when of large dimensions, is, if substantially fixed, quite as good, when of the same material, and much cheaper than wrought-iron. Where the shaft is exposed to impact, or any irregularity of force, wrought-iron has the superiority; but in other cases, when the castings are sound and good, cast-iron may be employed with perfect safety.

The dimensions required for a shaft, transmitting any given force, will depend on the resistance of the material of which it is composed. Consequently, the selection of material must be determined by the necessity for strength. Shafts may be considered as subject to two forces: a force producing simple flexure, arising from their own weight, the weight of the wheels and pulleys, and the strain of the belts; and a twisting force or torsion, arising from the power transmitted. If the flexure be great, the brasses will be much worn, vibration becomes considerable, and the disintegration of the machinery goes on in an accelerating ratio; it is therefore necessary to proportion shafting to the simple weight and direct transverse strain it has to sustain, so as to reduce the flexure within exceedingly narrow limits. In addition to this, the shafting, having to transmit a torsive force, must at least be capable of transmitting it without danger of rupture. In long and light shafting the tendency to flexure is usually greater than that to rupture by torsion;

the former consideration will therefore determine the size of the shaft. In short axles, etc., the danger from flexure almost disappears, and the strength of the shaft is determined by its resistance to torsion only. In all cases both conditions must be complied with, if security and permanence are to be obtained.

2. *Transverse Strain.*

Resistance to rupture. The general formula for resistance to rupture, in the case of a bar or beam supported at each end and loaded in the centre, is

$$W = \frac{a\,d\,c}{l} \ldots (1),$$

where W is the load in the centre, a the area of a section of the bar, perpendicular to the length; d the depth of the bar, and l its length. In this case c is derived from experiment, and is constant for similar bars or beams.

For rectangular bars this formula becomes,

$$W = \frac{c\,b\,d^2}{l} \ldots (2).$$

where b is the breadth and d the depth.

The value of c, for rectangular bars found by Mr. Barlow, for various materials, is given in the following table. In applying these numbers to calculations, it must be remembered that a and d are to be taken in inches, and l in feet; then c the centre breaking-weight, is found in lbs.

When the beam is supported at one end and loaded at the other, the formula is

$$w = \frac{c b d^2}{2 l} \ldots (3).$$

Value of c for different Materials.

	lbs.
English Malleable Iron	2050
Cast Iron	2548
Oak	400
Canadian Oak	588
Ash	675
Pitch Pine	544
Red Pine	447
Riga Fir	376
Mar Forest Fir	415
Larch	280

In my own experiments* I found the value of c for cast-iron to range from 1606 to 2615, the mean value being about 2050, as given above for malleable iron. Wrought-iron ranges from the value given above to 3000 lbs.

For cylindrical shafts supported horizontally the ultimate resistance to rupture is about

$$w = \frac{15000 \, d^3}{l} \text{ for wrought-iron,}$$

$$= \frac{19000 \, d^3}{l} \text{ for cast-iron,}$$

where w is the centre-breaking weight in lbs., d

---

* *On the Application of Cast and Wrought-Iron to Building Purposes,* p. 74, et seq.

the diameter, and l the length between supports in inches, the shaft being supported at the ends and loaded in the middle.

If the cylindrical shaft be loaded at one end and supported at the other, these formulæ become

$$w = \frac{7500\, d^3}{l} \text{ for wrought-iron,}$$

$$= \frac{9500\, d^3}{l} \text{ for cast-iron.}$$

If a beam be uniformly loaded over its entire length it will sustain twice the load that would break it if placed at the centre.

If the load be placed at any point intermediate between the centre and the ends, the breaking weight may be found by the following rule:—Divide four times the product of the distance in feet, of the weight from each bearing, by the whole distance in feet, and the quotient may be substituted for l in the formulæ above. That is, if x and y be its distances in feet from the two bearings respectively;

$$l = \frac{4\,x\,y}{(x+y)}.$$

From these rules the strength of shafts may be calculated, in all the cases of ordinary practice, where the tendency to transverse fracture has to be guarded against, making the actual strength at least five to ten times the strain to be carried. In shafting, however, it is not usually the transverse rupture, but the flexure produced by lateral stress, which limits the size of the shaft;—stiffness in fact becomes, in these cases, a more important element than strength.

The following formula has been given for the deflection of bars or beams loaded at the centre and supported at the ends:—

Let, d be the depth in inches;
b the breadth in inches;
L the length between supports in feet;
W the load in lbs.;
δ the deflection at the centre in inches;
M the modulus of elasticity;

then:— $W = \dfrac{M b d^3 \delta}{432 L^3}$; and if $b = d$,—

$$d^4 = \frac{432 L^3 W}{M \delta} \text{ or } d = \sqrt[4]{\frac{432 L^3 W}{M \delta}} \ldots\ldots(4).$$

Or, in words, multiply the product of the load in lbs., and the cube of the length in feet, by 432, and divide by the product of the modulus of elasticity and the deflection assumed in inches; the fourth root of the quotient will be the side of a shaft or square section which would deflect δ inches with a weight of W lbs. placed at its centre.*

---

* *Engineer and Machinist's Assistant*, p. 135, from which formulæ (4), (5), (6), to (11), and (23), in their present convenient form for practical use, have been quoted. The fundamental formula, however, is due to Young (*Nat.*

STRENGTH AND PROPORTIONS OF SHAFTS. 183

The following table gives the values of the modulus of elasticity for various materials :—

Modulus of elasticity in lbs.

Cast-iron........13,000,000 to 22,907,000
"   "   mean................17,000,000
Malleable iron...24,000,000 to 29,000,000
Steel............29,000,000 to 42,000,000
Brass.........................8,930,000
Tin..........................4,608,000
Ash..........................1,600,000
Beech........................1,353,600
Red pine, mean...............1,700,000
Spruce, mean.................1,600,000
Larch..............900,000 to 1,360,000
English oak......1,200,000 to 1,750,000
American oak.................2,150,000

For a cylindrical shaft, the same formula will apply with another constant. I am not aware that this has been experimentally ascertained, but it has been given approximately as 734. Hence, for cylindrical shafts,

$$d^4 = \frac{734 \, L^3 W}{M \delta} \quad \text{or} \quad d = \sqrt[4]{\frac{734 \, L^3 W}{M \delta}} \quad \ldots\ldots(5).$$

In the work just quoted, these formulæ have been simplified, by fixing a maximum value for δ, the deflection. The writer assumes that, with shafting, the deflection ought never to exceed $\frac{1}{100}$ of an inch for every foot length of the shaft. Substituting this value, and also the numerical

---

Philos., vol. ii., art. 326), and to Tredgold (Strength of Cast-Iron, p. 208).

value of the modulus of elasticity, he obtains the following formulæ:—

1. *For wood,*—taking M generally = 1,500,000, and $\delta = \dfrac{L}{100}$ inches.

Then, for square shafts, d being the depth of the side of the square—

$$d^4 = \dfrac{L^2 W}{35} \ldots (6).$$

And for round shafts, d being the diameter in inches—

$$d^4 = \dfrac{L^2 W}{20} \ldots (7).$$

2. *For cast-iron*—taking M = 18,000,000 lbs. and L as before—

For square section, $d^4 = \dfrac{L^2 W}{412} \ldots (8).$

For round section, $d^4 = \dfrac{L^2 W}{240} \ldots (9).$

3. *For wrought-iron*—taking M = 24,500,000 lbs. and δ as before—

For square section, $d^4 = \dfrac{L^2 W}{567} \ldots (10).$

For round section, $d^4 = \dfrac{L^2 W}{334} \ldots (11).$

By transposition, the formulæ given above become,—

STRENGTH AND PROPORTIONS OF SHAFTS.

For wood—

$$\text{Square section,} \quad L = \sqrt{\frac{35\,d^4}{W}} \quad \ldots (12).$$

$$W = \frac{35\,d^4}{L^2}$$

$$\text{Round section,} \quad L = \sqrt{\frac{20\,d^4}{W}} \quad \ldots (13).$$

$$W = \frac{20\,d^4}{L^2} \quad \ldots (14).$$

For cast-iron—

$$\text{Square section,} \quad L = \sqrt{\frac{412\,d^4}{W}} \quad \ldots (15).$$

$$W = \frac{412\,d^4}{L^2} \quad \ldots (16).$$

$$\text{Round section,} \quad L = \sqrt{\frac{240\,d^4}{W}} \quad \ldots (17).$$

$$W = \frac{240\,d^4}{L^2} \quad \ldots (18).$$

For wrought iron—

$$\text{Square section,} \quad L = \sqrt{\frac{567\,d^4}{W}} \quad \ldots (19).$$

$$W = \frac{567\,d^4}{L^2} \quad \ldots (20).$$

$$\text{Round section,} \quad L = \sqrt{\frac{334\,d^4}{W}} \quad \ldots (21).$$

$$W = \frac{334\,d^4}{L^2} \quad \ldots (22).$$

When the weight is uniformly distributed over the length of the shaft, the general formula is

$$d^4 = \frac{270\, L^3\, W}{M\, \delta} \text{ or } d = \sqrt[4]{\frac{270\, L^3\, W}{M\, \delta}} \ldots (23).$$

Substituting in this equation the same values of M and δ as before, we obtain the following formulæ:—

For wood— $\quad d^4 = \dfrac{L^3\, W}{56}$ for square shafts.

$\quad\quad\quad\quad\quad\quad d^4 = \dfrac{L^3\, W}{32}$ for round shafts.

For cast-iron— $d^4 = \dfrac{L^3\, W}{666}$ for square shafts.

$\quad\quad\quad\quad\quad\quad d^4 = \dfrac{L^3\, W}{383}$ for round shafts.

For wrought-iron $d^4 = \dfrac{L^3\, W}{907}$ for square shafts.

$\quad\quad\quad\quad\quad\quad d^4 = \dfrac{L^3\, W}{521}$ for round shafts.

The following tables for cast and wrought-iron round shafting, are calculated from the formulæ (9) and (11) for weights placed at the centre of a shaft supported at each end. In using them for cases in which the weight is distributed along its length, as in the case of the weight of the shaft itself, it must be remembered that a distributed weight produces ⅝ths of the deflection of the same weight placed at the centre.

STRENGTH AND PROPORTIONS OF SHAFTS. 187

TABLE 1.—RESISTANCE TO FLEXURE. WEIGHTS PRODUCING A DEFLECTION OF 1/40TH OF THE LENGTH IN CAST-IRON CYLINDRICAL SHAFTS.

Length between supports in feet.	Diameter of Shaft in Inches.																	
	1	1½	2	2½	3	3½	4	4½	5	5½	6	7	8	9	10	12	14	16	18	20
	lbs.	lbs.	lbs	lbs.	lbs.	lbs.	lbs.	lbs.	lbs.	lbs.	lbs.	lbs.	lbs.	lbs.	lbs.	tons.	tons.	tons.	tons.	tons.
5	9·6	49	164	375	778	1,441	2,458	3,937	6,000	8,785	12,442	23,050	39,322	62,986	96,000	88·9	164·6	280·9	449·9	685·7
6	6·7	34	107	261	640	1,000	1,707	2,734	4,166	6,100	8,640	16,007	27,307	43,741	66,660	61·7	114·3	195·1	312·4	476·2
7	4·9	25	78	191	397	735	1,254	2,008	3,061	4,482	6,348	11,760	20,062	31,673	48,980	45·3	84·0	143·3	229·5	349·8
8	3·7	19	60	146	304	563	960	1,538	2,344	3,431	4,860	9,004	15,360	24,604	37,500	34·7	64·3	109·7	175·7	267·9
9		15	47	116	240	446	768	1,215	1,852	2,711	3,840	7,115	12,136	19,440	29,630	27·6	50·8	86·7	138·0	211·6
10		12	38	94	194	360	615	984	1,500	2,196	3,110	5,763	9,830	15,747	24,000	22·2	41·2	70·2	112·5	171·4
11		10	32	77	160	298	508	814	1,240	1,815	2,571	4,763	8,124	13,014	19,834	18·4	34·0	58·0	93·0	141·7
12		8	27	65	135	251	427	683	1,042	1,625	2,160	4,002	6,827	10,935	16,667	15·4	28·6	48·8	78·1	119·1
13			23	55	115	213	364	583	888	1,299	1,840	3,410	5,817	9,317	14,201	13·1	24·4	41·5	66·6	101·4
14			19	48	99	184	314	502	765	1,121	1,587	2,940	5,016	8,034	12,245	11·3	21·0	35·8	57·4	87·4
15			17	42	86	160	273	437	666	976	1,383	2,661	4,369	7,000	10,667	9·9	18·3	31·2	50·0	76·2
16			15	37	76	141	240	384	586	858	1,215	2,251	3,840	6,151	9,354	8·7	16·1	27·4	43·9	66·9
17			13	33	67	125	213	341	519	760	1,076	1,994	3,401	5,448	8,305	7·7	14·2	24·3	39·0	59·3
18			12	29	63	112	192	309	471	686	960	1,779	3,034	4,860	7,407	6·9	12·7	21·7	34·7	52·9
19			11	28	54	100	173	277	423	615	868	1,606	2,738	4,397	6,700	6·2	11·4	19·6	31·3	47·9
20			10	23	48	90	154	246	375	550	778	1,441	2,458	3,937	6,000	5·5	10·3	17·5	28·1	42·9

TABLE 2.—RESISTANCE TO FLEXURE. WEIGHTS PRODUCING A DEFLECTION OF $\frac{1}{1000}$TH OF THE LENGTH IN WROUGHT-IRON CYLINDRICAL SHAFTS.

Length between bearings in feet.	Diameter of Shaft in Inches.																	
	1	1½	2	2½	3	3½	4	4½	5	6	7	8	9	10	12	14	16	18	20
	lbs.	lbs.	lbs.	lbs.	lbs.	lbs.	lbs.	lbs.	lbs.	lbs.	lbs.	lbs.	lbs.	lbs.	tons.	tons.	tons.	tons.	tons.
5	13·4	68	214	522	1,082	2,006	3,420	5,478	8,350	17,314	32,078	54,724	87,664	133,600	123·6	229·1	391·0	626·0	954
6	9·3	47	148	363	751	1,392	2,375	3,804	5,799	12,024	22,277	38,003	60,871	92,780	86·0	159·1	272·0	435·0	663
7	6·8	35	109	267	552	1,023	1,745	2,795	4,260	8,834	16,367	27,920	44,721	68,163	63·1	116·9	200·0	310·4	487
8	5·2	26	84	204	423	783	1,336	2,140	3,262	6,764	12,531	21,376	34,240	52,188	48·3	89·5	153·0	244·6	373
9		21	66	161	334	619	1,055	1,691	2,577	5,344	9,900	16,800	27,053	41,236	38·1	70·7	121	193·2	295
10		17	53	130	270	501	855	1,370	2,087	4,329	8,020	13,681	21,913	33,400	30·9	57·3	97·7	156·5	231
11		14	44	108	223	414	707	1,132	1,725	3,577	6,628	11,306	18,110	27,603	25·5	47·3	80·7	129·4	197
12		12	37	91	188	348	594	952	1,450	3,006	5,669	9,500	15,218	23,195	21·5	39·8	67·0	108·7	166
13		10	31	77	160	297	507	810	1,236	2,561	4,745	8,095	12,066	19,763	18·3	33·9	57·8	92·6	141
14			27	68	140	258	436	699	1,065	2,208	4,091	6,980	11,180	17,041	15·8	29·2	49·9	79·9	122
15			24	60	138	223	380	609	929	1,924	3,564	6,080	9,740	14,844	13·7	25·5	43·4	69·6	106
16			21	51	122	196	334	535	816	1,691	3,133	5,344	8,560	13,047	12·0	22·4	38·1	61·1	103·2
17			19	46	106	173	290	474	722	1,498	2,775	4,734	7,582	11,557	10·7	19·8	33·8	54·2	82·5
18			17	42	96	157	268	430	655	1,359	2,518	4,296	6,880	10,488	9·7	18·0	30·6	48·3	74·8
19			15	37	84	141	240	386	589	1,221	2,261	3,858	6,179	9,410	8·7	16·1	27·5	43·5	67·2
20			13	33	76	125	213	342	522	1,082	2,005	3,420	5,478	8,350	7·7	14·3	24·4	39·1	59·8

STRENGTH AND PROPORTIONS OF SHAFTS.

From the foregoing it will be seen, that the weights given in the tables are correct indications of the load required in the centre to produce a deflection of the $\frac{1}{1200}$ of the length of the shaft.* This fraction is not however the universal standard among millwrights; on the contrary, there appears to be no recognized standard in practice, by which the deflection from a given weight can be ascertained, and although $\frac{1}{1200}$ may, in many cases, give a larger area with increased weight, in shafts that are not heavily loaded in the middle, nevertheless it is important that the shafts, when loaded as above, should not bend more than $\frac{1}{1200}$ of their length. In cases where the load is light and equally distributed, lighter and smaller shafts would suffice.

The following tables give the deflection of cylindrical shafts with their own weight:—

TABLE 3.—DEFLECTION ARISING FROM THE WEIGHT OF THE SHAFT. CAST-IRON CYLINDRICAL SHAFTS.

Length between bearings in feet.	Diameter of Shaft in Inches.								
	1	2	4	6	8	10	12	14	16
ins.	ins.	ins.	ins.	ins.	ins.	ins.	ins.	ins.	ins.
5	·004	·001	·000	·000	·000	·000	·000	·000	·000
10	·067	·017	·004	·002	·001	·001	·001	·000	·000
15	·338	·085	·021	·009	·005	·003	·002	·002	·001
20	1·067	·267	·067	·029	·017	·011	·007	·005	·004
25	2·603	·651	·163	·073	·041	·026	·018	·013	·010

* This standard is the one assumed by Tredgold (Strength of Cast-Iron, p. 210).

TABLE 4.—DEFLECTION ARISING FROM THE WEIGHT OF THE SHAFT. WROUGHT-IRON CYLINDRICAL SHAFTS.

Length between bearings in feet.	Diameter of Shaft in Inches.								
	1	2	4	6	8	10	12	14	16
ins.	ins.	ins.	ins.	ins.	ins.	ins.	ins.	ins.	ins.
5	·003	·001	·000	·000	·000	·000	·000	·000	·000
10	·050	·013	·003	·001	·001	·001	·000	·000	·000
15	·256	·064	·016	·007	·004	·003	·002	·001	·001
20	·808	·202	·051	·022	·013	·008	·005	·004	·003
25	1·972	·493	·123	·055	·031	·020	·013	·010	·008

The above tables clearly indicate the deflection of shafts of different lengths by their own weight, and will be a guide to the millwright in calculating the distance of the bearings between which they revolve. It is important in shafting, when extended in long ranges, that there should not be any serious deflection, either from the weight of the shaft, or lateral stress; I have always found that a stiff shaft, although heavier in itself, is lighter to retain in motion than a smaller one which bends to the strain.

3. *Torsion.*

In addition to the lateral flexure from transverse forces, shafting is subjected to a wrenching or twisting, from the power transmitted acting tangentially to its circumference. This causes one end of the shaft to revolve in relation to the

STRENGTH AND PROPORTIONS OF SHAFTS.

other end, through a smaller or greater angle, known as the angle of torsion, and if sufficient force be applied, this angle increases till the resistance of the material is overcome, and the shaft gives way.

Coulomb laid the basis of our knowledge of the resistance to torsion of cylindrical bodies, and he verified his theoretical deductions by admirably-contrived experiments, on a small scale. He showed that in wires where the diameter is small in relation to the length, the angles of torsion are in proportion to the length, and reciprocally proportional to the moment of inertia of the base of the cylinder in relation to its centre. He also discovered that each wire acquired a permanently acceleration-varying torsion, according to the degree in which it departed from its primitive position, and that these permanent torsions have no fixed relation to the temporary torsions, coexisting with the application of the moving force. With the same wire he found the torsion to be in proportion to the force applied; with the same length and force inversely as the fourth power of the diameter.

These deductions are expressed by the following formula :—

$$\theta = \frac{2R}{\pi G} \times \frac{wl}{r^4}$$

where θ is the angle of torsion, r the radius, and l the length of the wire, R the leverage at which the weight W acts, and G the modulus of torsion for the material; being about ⅔ths of the modulus of elasticity.

In 1829 a paper was communicated to the Royal Society by Mr. Bevan, containing experimental determinations of the modulus of torsion for a large number of substances, of which the most important are given below.

Let δ be the deflection of a prismatic shaft of a given length l when strained by a given force w in lbs., acting at right angles to the axes of the prism and at a leverage r; let d be the side of the square section of the shaft, l, r, δ, d being in inchs.

$$\delta = \frac{r^3 l w}{d^4 \text{T}}$$

where T is the modulus of elasticity in the following table.

If the transverse section of the prism be a parallelogram, let b be the breadth and d the depth, then Mr. Bevan gives the formula—

$$\delta = \frac{(d+b) l r^3 w}{2 b d^3 \text{T}}$$

If the torsion be required in degrees (Δ), then let $\rho = 57\cdot 29578$,

$$\Delta = \frac{r \rho l w}{d^4 \text{T}}, \text{ for square shafts.}$$

For example,

$$\Delta = \frac{r l w}{31000\, d^4} \text{ for wrought-iron and steel,}$$

$$= \frac{rlw}{16600\,d^4}, \text{ for cast-iron.}$$

A very careful experimental study of the effect of torsion on various materials has been made by Mr. M. G. Wertheim, and was presented to the Académie des Sciences in 1855. The general results at which he has arrived may be stated as follows:—

1. The total angle of torsion consists of two parts, of which one is purely temporary, whilst the other persists after the force has ceased to act. It is not possible to assign the limit at which the permanent torsion begins to be sensible, nor has it any fixed relation to the temporary torsion; it augments at first very slowly, afterward more rapidly, till the bar breaks.*

---

* We have many practical instances of this tendency to rupture which at first appear only temporary, but a continuation of the same action, particularly in long ranges of shafts, in process of time, developes itself in the form of a permanent deterioration which ultimately leads to fracture. This was strikingly exemplified in a range of shafts, 220 feet long, tapering from three inches diameter at the driving end, to two inches diameter at the other.

The work done by these shafts was uniform throughout, but it was soon found that the shaft had made nearly 1·16 revolutions at the driven end of the room, before it began to move at the other. The result was a continued series of jerks or accelerated and retarded motion, injurious to the machinery, and destructive to the work it had to perform. It was, moreover, injurious to the shafts, particularly in the middle, where the twist was severely felt, and

TABLE 5.—VALUES OF MODULUS OF TORSION ACCORDING TO MR. BEVAN.

Material.	Specific gravity.	Modulus of torsion. (T).	
		lbs.	
Ash	—	20,300	
Beech	—	21,243	
Elm	—	13,500	
Scotch fir	—	13,700	
Hornbeam	·86	26,400	
Larch	·58	18,967	
English oak	—	20,000	
Memel pine	—	15,000	
American pine	—	14,750	
Teak	—	16,800	Old and partially decayed.
Teak, African	—	27,300	
Iron, English wrought	—	1,775,000	(Mean.)
Steel	—	1,753,000	(Mean.)
Iron (cylindrical)	—	1,910,000	
" "	—	1,700,000	
" (square)	—	1,617,000	
" "	—	1,667,000	
" "	—	1,951,000	
Cast-iron	—	940,000	
" "	—	963,000	
" "	—	952,000	
" "	—	951,600	(Mean.)
Bell metal	—	818,000	

2. The temporary angles are not rigorously proportional to the moments of the forces applied.

3. The mean angles of torsion are not rigorously proportional to the length of the bar, increasing,

---

would have led to rupture, but from the circumstance that they had to be renewed with a stiffer and stronger range.

although very slightly, in proportion to the length, as the bars are made shorter.

4. The interior cavity of all hollow homogeneous bodies diminish by torsion, and this diminution is proportional to the length and to the square of the angle of torsion for unity of length.

5. For cylindrical bodies Mr. Wertheim gives the following formulæ :—

Let ψ be the mean temporary angle of torsion, for

$p = 1$ kilogramme, and
$l = 1$ mètre;

$p =$ the sum* of the two weights producing torsion and constituting a couple in kilogrammes;

$R =$ the leverage at which the weight p acts;

$l =$ length of the bar subject to torsion, in millimètres;

$r =$ the exterior radius of the section of the bar, in Millimètres;

$r_1 =$ the interior radius of hollow bars, in millimètres;

$E =$ the modulus of elasticity of the material obtained from experiments on tension.

---

* In Mr. Wertheim's experiments equal weights, acting in opposite directions at the same leverage, were hung one on each side of the bar, subjected to torsion.

Then for solid bars:—*

$$\Psi = \frac{16}{3} \cdot \frac{180}{\pi^2} \cdot \frac{pR}{E} \cdot \frac{l}{r^4}$$

and for hollow cylinders

$$\Psi = \frac{16}{3} \cdot \frac{180}{\pi^2} \cdot \frac{pR}{E} \cdot \frac{l}{r^4 - r'^4}$$

In the following experiments, $p = 1$ kilogramme, $R = 247·5$ millimètres, $l = 1000$ millimètres.

RESUMÉ OF EXPERIMENTS ON CYLINDERS OF CIRCULAR SECTION.

	Material.	Radius r.	Coefficient of elasticity, E.	Mean angle of torsion.	
				By formula.	By experiment.
		mm.		° ′ ″	° ′ ″
1	Iron	8·220	17,805	0 17 46·1	0 17 52·1
2	Iron	5·501	"	1 28 0·8	1 26 31·3
3	Cast steel	5·055	19,542	1 53 12·0	1 51 13·4
4	Copper	5·031	9,395	3 59 59·1	3 54 6·0
5	Glass	3·535	6,200	24 51 56·0	24 15 34·7
6	Glass	3·4225	"	28 18 2·0	28 30 14·0

The accordance, in these tables, between the formulæ and the experiments is very satisfactory, especially considering that the value of E cannot be determined with perfect accuracy. The errors do not generally exceed $\frac{1}{50}$th, and the observed

---

* The above formulæ may be used with English measures, E being taken from English tables, if p be given in lbs. and r, l, and R in inches.

angles are smaller than those found by calculation except in the case of the cylinders 9, 53, and 54.

Résumé of Experiments on the Torsion of Hollow Cylinders of Copper.

	External radius (r).	Internal radius (r1).	Coefficient of elasticity from tension (E).	Angle of torsion (Ψ).	
				By formula.	By experiment.
				° ′ ″	° ′ ″
53	11,525	10,021	10,917	0 17 30·2	0 20 0·6
54	7,082	4,955	10.444	1 12 18·3	1 16 52·9
55	5,047	30,315	10,276	4 9 4·0	4 6 54·7
7	5,602	24,665	9,665	2 37 40·4	2 33 38·2
8	45,605	2,478	9,855	6 11 10·3	6 0 53·8
9	36,955	2,471	10,645	15 9 14·4	15 42 37·3

For bars of elliptical section M. Wertheim has deduced the formula

$$\Psi = \frac{8}{3} \cdot \frac{180}{\pi^2} \cdot \frac{p\,R}{E} \cdot \frac{l(c_1^2 + c_2^2)}{c_1^2\, c_2^2}$$

where c_1 and c_2 are the two semiaxes of the ellipse, the other letters remaining as before.

Résumé of Experiments on the Torsion of Elliptical Bars.

	Material.	Semiaxes.		Coefficient of elasticity by tension (E.)	Mean angle of torsion (Ψ).	
		c_1	c_2		By formula.	By exper't.
		mm.	mm.		° ′ ″	° ′ ″
11	Cast steel...	7,105	3,697	19,085	2 13 56·7	2 10 55·4
12	" ...	9,900	25,075	"	4 18 0·1	4 13 18·2
13	Copper.......	7,062	3,669	9,634	4 32 56·7	4 30 41·2
14	"	9,875	2,496	"	8 38 11·2	8 54 33·9

For bars of rectangular section the formula becomes

$$\psi = \frac{180}{\pi^2} \cdot \frac{1}{2} \cdot \frac{p\,\mathrm{R}}{\mathrm{E}} \cdot \frac{l(a^2 + b^2)}{a^3 b^3}$$

But it is necessary to apply a coefficient of correction c to the calculated angle such that if ψ^1 be the calculated angle of torsion, and ψ^2 the angle found by experiment, then $c = \dfrac{\psi_1}{\psi^2}$. This coefficient varies with the ratio $\dfrac{a}{b}$ of the sides of the bar. Thus when $l = 500$ millimètres, and the section was 36 milimètres square.

$\dfrac{a}{b}$	1	2	4	8
Value of coefficient	0·8971	0·9617	0·9520	0·9878

It varies also with the ratio $\dfrac{l}{b}$ and with the moment of the couple $p\,\mathrm{R}$.

For the ultimate resistance of cylindrical shafts to rupture by torsion, Professor W. J. M. Rankine gives the following formula:[*]

Let l denote the length in inches of the lever, such as a crank, at the end of which a wrenching or twisting force is applied to an axle. Let w be the working load in pounds, multiplied by a suit-

---

[*] Manual of Applied Mechanics, p. 355. Manual of Steam Engine, p. 78.

able factor of safety (usually six); then
$$W l = M$$
is the wrenching moment in inch pounds.

For a solid axle let h be its diameter; then
$$M = \frac{f h^3}{5\cdot 1} \text{ and } h = \sqrt[3]{\frac{5\cdot 1\, M}{f}}.$$

For a hollow axle let h_1 be the external, and h_0 the internal diameter in inches; then
$$M = \frac{f(h_1^4 - h_0^4)}{5\cdot 1\, h_1} = \frac{f h_1^3}{5\cdot 1} \cdot \left(1 - \frac{h_0^4}{h_1^4}\right)$$
$$\text{and } h_1 = \sqrt[3]{\left\{ \frac{5\cdot 1\, M}{f\left(1 - \frac{h_0^4}{h_1^4}\right)} \right\}}$$

The values of the modulus of wrenching f are—

for cast-iron   about   30000,
for wrought-iron   "   54000,

and taking six as the factor of safety, if we put the working moment of torsion in the formulæ instead of the wrenching moment, we may put instead of f

for cast iron   5000,
for wrought iron 9000.

Hence we get for W, the working stress, with solid shafts,
$$W_1 = \frac{5000\, h^3}{5.1\, l} = \frac{980\, h^3}{l} \text{ for cast-iron } \dots (2.)$$
$$= \frac{9000\, h^3}{5\cdot 1\, l} = \frac{1765\, h^3}{l} \text{ for wrought-iron}..(3.)$$

On this principle I have calculated the following tables (pages 200, 201,) giving the safe mo-

TABLE 6.—SAFE WORKING TORSION FOR CAST-IRON SHAFTS.

Diameter of Shaft in Inches.	Working moment of torsion in inch lbs. (W₁.)	Working Stress in lbs. at the following Radii, or W₁.									
		6 ins.	9 ins.	12 ins.	16 ins.	18 ins.	21 ins.	24 ins.	27 ins.	30 ins.	42 ins.	60 ins.
1	980	163	109	81	64	54	46	41	36	32	23	16
1½	3,309	552	368	276	220	184	158	138	122	110	79	55
2	7,843	1,307	871	653	522	435	372	326	290	261	186	131
2½	15,320	2,553	1,702	1,276	1,020	851	730	638	567	510	365	256
3	26,470	4,411	2,941	2,206	1,764	1,470	1,260	1,103	980	882	630	441
3½	42,030	7,005	4,670	3,502	2,802	2,335	2,002	1,751	1,556	1,401	1,001	701
4	62,740	10,457	6,971	5,228	4,182	3,485	2,988	2,614	2,323	2,091	1,494	1,046
4½	89,350	14,891	9,928	7,446	5,956	4,964	4,254	3,722	3,309	2,978	2,127	1,489
5	122,650	20,425	13,617	10,212	8,170	6,809	5,834	5,106	4,539	4,085	2,917	2,043
5½	163,110	27,185	18,123	13,592	10,874	9,061	7,766	6,796	6,041	5,437	3,883	2,718
6	211,700	35,283	23,528	17,646	14,116	11,764	10,084	8,823	7,842	7,058	5,042	3,529
6½	269,240	44,873	29,916	22,436	17,960	14,965	12,820	11,218	9,972	8,975	6,410	4,487
7	339,280	56,047	37,364	28,023	22,418	18,682	16,012	14,012	12,454	11,209	8,005	5,605
7½	413,600	68,933	45,956	34,466	27,572	22,977	19,696	17,233	15,318	13,786	9,848	6,893
8	501,970	83,661	55,774	41,830	33,464	27,887	23,904	20,916	18,591	16,732	11,952	8,366
8½	602,100	100,350	66,900	50,175	40,140	33,450	28,672	25,088	22,300	20,070	14,336	10,035
9	714,720	119,120	79,413	59,560	47,648	39,706	34,034	29,780	26,471	23,824	17,017	11,912
9½	840,570	140,095	93,396	70,047	56,038	46,698	40,026	35,024	31,132	28,019	20,013	14,010
10	980,000	163,333	108,888	81,666	65,332	54,444	46,666	40,833	36,296	32,666	23,333	16,333
11	1,304,900	217,483	144,988	108,741| 86,994	72,494	62,138	54,371	48,329	43,497	31,069	21,748
12	1,694,100	282,350	188,233	141,176	112,940	94,116	80,672	70,587	62,744	56,470	40,336	28,236
13	2,154,000	359,000	239,333	179,500	143,600	119,666	102,570	89,750	79,777	71,800	51,285	35,900
14	2,690,200	448,366	298,911	224,183	179,348	149,455	128,104	112,091	99,637	89,673	64,052	44,837
15	3,308,900	551,466	367,644	275,733	220,586	183,772	157,562	137,866	122,614	110,293	78,781	55,147
16	4,015,700	669,283	445,200	334,641	267,712	222,900	191,224	167,321	148,400	133,856	95,612	66,928
17	4,816,800	802,800	535,200	401,400	321,120	267,600	229,370	200,700	178,400	160,560	114,686	80,280
18	5,718,000	953,000	635,333	476,500	381,200	317,666	272,286	238,250	211,777	190,600	136,143	95,300
19	6,724,800	1,120,766	747,177	560,383	448,306	373,688	320,218	280,191	249,059	224,163	160,109	112,077
20	7,843,000	1,307,166	871,444	653,583	522,866	435,722	375,762	328,791	290,481	261,433	186,881	130,717

STRENGTH AND PROPORTIONS OF SHAFTS. 201

TABLE 7.—SAFE WORKING TORSION FOR WROUGHT-IRON SHAFTS.

Diameter of Shaft in inches.	(W₁ L) Working moment of torsion in inch lbs.	Working Stress in lbs. at the following Radii.										
		6 ins.	9 ins.	12 ins.	15 ins.	18 ins.	21 ins.	24 ins.	27 ins.	30 ins.	42 ins.	60 ins.
1	1,765	294	196	147	118	98	84	73	65	59	42	29
1½	5,956	923	662	496	396	331	284	248	221	198	142	99
2	14,120	2,353	1,569	1,176	942	784	672	588	523	471	336	235
2½	27,570	4,595	3,063	2,297	1,838	1,531	1,312	1,148	1,021	919	656	460
3	47,650	7,942	5,294	3,971	3,176	2,647	2,270	1,985	1,764	1,588	1,135	794
3½	75,060	12,510	8,406	6,305	5,044	4,203	3,602	3,152	2,802	2,522	1,801	1,261
4	112,940	18,823	12,549	9,411	7,530	6,274	5,378	4,705	4,183	3,765	2,689	1,882
4½	160,806	26,801	17,867	13,400	10,720	8,933	7,658	6,700	5,956	5,360	3,829	2,680
5	220,590	36,765	24,510	18,382	14,706	12,256	10,504	9,191	8,170	7,353	5,252	3,676
5½	293,600	48,933	32,622	24,466	19,572	16,311	13,980	12,233	10,874	9,786	6,990	4,893
6	381,190	63,530	42,353	31,765	25,412	21,176	18,152	15,882	14,117	12,706	9,076	6,353
6½	484,630	80,771	53,848	40,385	32,308	26,924	23,076	20,192	17,949	16,154	11,539	8,077
7	605,290	100,881	67,254	50,440	40,352	33,627	28,822	25,220	22,418	20,176	14,411	10,088
7½	744,480	124,080	82,720	62,040	49,632	41,360	35,452	31,020	27,673	24,816	17,726	12,408
8	903,530	150,588	100,392	75,294	60,234	50,196	43,024	37,647	33,464	30,117	21,512	15,059
8½	1,083,600	180,633	120,422	90,316	72,254	60,211	51,610	45,168	40,140	36,127	25,805	18,063
9	1,286,500	214,416	142,944	107,208	85,766	71,472	61,262	53,604	47,648	42,883	30,631	21,442
9½	1,513,000	252,166	168,111	126,083	100,866	84,055	72,048	63,041	56,037	50,433	36,024	25,217
10	1,765,000	294,166	196,111	147,083	117,666	98,055	84,048	73,541	65,370	58,833	42,024	29,417
11	2,348,400	391,466	260,966	195,733	156,586	130,483	111,848	97,866	86,988	78,293	55,924	39,147
12	3,049,400	508,233	338,822	254,116	203,292	169,411	139,496	127,058	112,941	101,646	69,748	50,823
13	3,877,100	646,123	430,788	323,091	258,474	215,394	164,624	161,545	143,596	129,237	92,312	64,618
14	4,842,300	807,050	538,033	403,525	322,620	269,016	230,586	201,762	179,344	161,410	115,293	80,705
15	5,955,800	992,633	661,755	496,316	397,054	330,677	283,610	248,158	220,585	198,527	141,805	99,263
16	7,228,200	1,204,700	803,133	602,350	481,680	401,566	344,200	301,175	267,711	240,940	172,100	120,470
17	8,670,100	1,445,016	963,366	722,508	578,006	481,683	412,862	361,254	321,122	289,003	206,431	144,501
18	10,292,000	1,715,333	1,143,556	857,666	686,132	571,777	490,096	428,833	381,185	343,066	245,048	171,533
19	12,104,000	2,017,333	1,344,888	1,008,666	806,932	672,444	576,380	504,333	448,296	403,466	288,190	201,733
20	14,120,000	2,353,333	1,568,888	1,176,666	941,332	784,444	672,380	588,333	522,962	470,666	336,190	235,333

ment of torsion for cylindrical cast and wrought-iron shafts, and also the working stress to which they may be subjected at the circumference of pullies or wheels of various diameters. In cases where the horses' power transmitted by a shaft is given instead of the stress, the latter may be found by the table on page 172.

The greatest angle of torsion, which it is safe to allow in a line of shafting, is determined by the extension of the material within the elastic limits. If $\frac{1}{1200}$th of the length be assumed as the maximum extension with the safe working load, then the shaft must be so proportioned that the angle of torsion is less than that given by the following formula:

$$\psi = \frac{2284\,L}{1000\,d} \quad \ldots(4.)$$

where L is the length of the shaft in feet, d its diameter in inches, and ψ the angle of torsion in degrees.

It is convenient to estimate the ultimate resistance of shafts to torsion, not only as a statical pressure acting at a leverage, but also in horses' power. Now the stress resulting from the transmission of power must evidently increase in proportion to the power, and decrease in proportion to the velocity. A shaft will transmit 100 horses' power at 80 revolutions a minute with no more stress than it would transmit 50 horses' power at

STRENGTH AND PROPORTIONS OF SHAFTS. 203

40 revolutions, or 25 horses' power at 20 revolutions. Hence the torsion varies as $\frac{H}{R}$, where H is the number of horses' power per minute, and R the number of revolutions per minute.

Buchanan's rules for the power transmitted by shafts are:—*

For fly-wheel shafts

$$d = \sqrt[3]{\left\{\frac{H}{R} \times 400\right\}}$$

For shafts of water-wheel gearing and other heavy work,

$$d = \sqrt[3]{\left\{\frac{H}{R} \times 200\right\}}$$

For shafts of ordinary mill gearing

$$d = \sqrt[3]{\left\{\frac{H}{R} \times 100\right\}}$$

An ordinary allowance for wrought-iron shafting in practice is

$$d = \sqrt[3]{\left\{\frac{H}{R} \times 250\right\}} \quad \ldots \quad (5.)$$

From the foregoing observations in regard to torsion, and the power of transmission of shafts at different velocities, it is a desideratum of much importance to the engineer, so to proportion shafts in relation to their lengths as well as velocities, as to be within the limits of sensible permanent

---

* These rules will be found in the second edition of Buchanan, at pages 328, *et seq.*

torsion and flexure,* and at the same time to increase the speeds in a given ratio to the velocities of the machine and the nature of the work it has to execute. In the above disquisition we have only given the law and the safe measure of torsion as regards length and area, but much must still depend on the calculation and judgment of the millwright and engineer; in its application to the character of the work they have to perform, and the resistances they have to overcome.

From formula (5.) the following table (page 205) has been calculated, giving the diameter necessary to transmit from 1 to 150 horses' power at from 10 to 1000 revolutions per minute.

4. *Velocity of Shafts.*

As the quality of the material employed for the construction of shafts enters largely into the calculation of their strength, so also the velocity at which they revolve becomes an important element in the calculation of the work transmitted by them. In all cases where machinery has to be driven at a high speed, it is advantageous and

---

* Although we speak of the limits of permanent torsion, we are not prepared to fix these limits, as we find that what produces a permanent set in any material, however minute a fraction it may be, will, in process of time, if continued, and often repeated, lead to fracture. This law applies to every description of strain or material, and we may therefore consider that there are limits to endurance, however distant that may be.

STRENGTH AND PROPORTIONS OF SHAFTS.

TABLE 8.—DIAMETER OF WROUGHT-IRON SHAFTING NECESSARY TO TRANSMIT WITH SAFETY VARIOUS AMOUNTS OF FORCE.

Horses power safely transmitted.	Diameter of Shaft in inches, with the following revolutions per minute.											
	10	25	50	75	100	150	200	250	300	400	500	1000
1	2·92	2·15	1·71	1·50	1·36	1·20	1·09	1·00	·94	·85	·79	·63
2	3·68	2·71	2·15	1·91	1·71	1·49	1·36	1·26	1·20	1·09	1·00	·79
3	4·22	3·11	2·47	2·15	1·95	1·71	1·54	1·44	1·36	1·23	1·14	·91
4	4·64	3·42	2·71	2·38	2·15	1·91	1·71	1·56	1·49	1·36	1·26	1·00
5	5·00	3·68	2·92	2·57	2·32	2·00	1·84	1·71	1·61	1·45	1·36	1·09
10	6·30	4·64	3·68	3·21	2·92	2·57	2·32	2·15	2·03	1·84	1·71	1·36
15	7·21	5·31	4·22	3·68	3·35	2·92	2·66	2·46	2·32	2·21	1·95	1·55
20	7·94	5·85	4·64	4·06	3·68	3·21	2·92	2·71	2·56	2·32	2·15	1·71
30	9·00	6·69	5·31	4·64	4·22	3·68	3·34	3·11	2·92	2·66	2·46	1·95
40	10·00	7·37	5·85	5·10	4·64	4·00	3·68	3·42	3·21	2·92	2·71	2·15
50	—	7·94	6·30	5·51	5·00	4·36	3·96	3·68	3·46	3·14	2·92	2·32
60	—	8·43	6·69	5·85	5·31	4·64	4·22	3·91	3·68	3·35	3·11	2·46
70	—	8·88	7·06	6·15	5·59	4·89	4·44	4·12	3·88	3·52	3·27	2·60
80	—	9·28	7·37	6·44	5·84	5·10	4·64	4·31	4·05	3·68	3·42	2·71
90	—	9·65	7·68	6·69	6·08	5·31	4·83	4·48	4·22	3·83	3·56	2·82
100	—	10·00	7·94	6·93	6·30	5·51	5·00	4·64	4·36	3·96	3·68	2·92
110	—	—	8·19	7·16	6·50	5·67	5·16	4·79	4·51	4·10	3·80	3·01
120	—	—	8·43	7·37	6·69	5·85	5·31	4·93	4·64	4·22	3·91	3·11
130	—	—	8·66	7·57	6·88	6·01	5·46	5·06	4·76	4·33	4·02	3·19
150	—	—	9·09	7·94	7·21	6·30	5·72	5·31	5·00	4·55	4·22	3·35

even essential to run the shafting at a proportionate velocity. If, for example, there are a series of machines running at five hundred revolutions per minute, it will be advisable to run the shafts at half that speed, by which means the following very important advantages will be gained:

There will be a great saving in the weight of the shafts, for with a slow motion of fifty revolutions per minute, fully three times the weight would be necessary to transmit the same power. There would also be a saving in original cost in the power absorbed, and in maintenance.

Shafts running at low velocities are cumbersome, heavy, and expensive to repair. They are costly in the first instance, and they block up the rooms of the mill with large drums and pullies, obstructing the light, which, in factories, is a consideration of very great importance.

At the commencement of the present century, mills were geared with ponderous shafts, such as those just described. They were generally of cast-iron, square, and badly coupled, and the power required to keep them in motion was in some cases almost equal to that required by the machinery they had to drive. In the present improved system, with light shafts accurately fitted and running at high velocities, the work which previously was absorbed in transmission is now conveyed to the machinery of the mill.

I may safely ascribe my own success in life

and that of my friend and late partner, Mr. James Lillie, to the saving of power effected by increasing threefold the velocity of the shafting in mills more than forty years ago. The introduction of light iron shafting not only enabled the manufacturer to effect a considerable saving in the original cost, but a still greater saving was effected in power, whilst it relieved the mills from the ponderous wooden drums and heavy shafting then in use, and established an entirely new system of operations in the machinery of transmission.

5. *Length of Journals.**

Another consideration of considerable importance to the smooth and safe working of shafting is the length of the journals. From a number of years' experience I have been led to believe, that with cast-iron, one and a half times the diameter of the shaft is the best proportion for the length of the bearing, and with wrought iron, one and three quarters the diameter. On the question of shafts revolving in the steps of plummer blocks and the proportions necessary to effect motion without danger of heating, it is essential (without entering largely into the laws of friction on bodies

---

* Rules for the diameters of gudgeons or journals for those cases in which they are calculated independently of the diameter of the shaft, are given in Mills and Millwork, vol. i. p. 116.

in contact) that we should ascertain from actual practice and long-tried experience the best form of journals of shafts adapted for that purpose. The lengths proportionate to the diameters have already been given, but we have yet to consider the dimensions of the journals of large shafts where they are small in comparison with the pressure or the weight they have to sustain. Let us, for example, take a fly-wheel shaft and the foot or toe of a line of vertical shaft extending to a height of six or seven stories in a mill filled with machinery, and we have the safe working pressure per square inch as indicated in the last column in the following table:—

Description of Shaft.	Length and diameter of shaft in inches.	No. of square inches in bearing.	Weight on bearing in lbs.	Weights in lbs. per square inch on bearing.
Fly-wheel shaft wrought-iron	18 × 14	252	45,024	178·21
Vertical shaft cast-iron	— × 11	95	23,061	242·70
Horizontal shaft cast-iron	15 × 10	150	6,000	40·00
Horizontal shaft wrought-iron	6 × 3	18	540	30·00
Ditto ditto ditto	2 × 4	8	160	20·00

From the above it will be seen that in fly-wheel shafts the pressure should never exceed 180 lbs. per square inch, and in that of the toes of vertical shafts 240 lbs. per square inch. Even with this latter pressure it is difficult to keep the shafts cool, and it requires the greatest possible care to keep them free from dust or any minute particles of sand or other sharp substances getting into the steps. The feet of vertical shafts also require the very best quality of gun metal for the shaft to run in, and fine limpid oil for lubrication to prevent the toe from cutting. It is, moreover, necessary for the shaft to fit well on the bottom of the step, and not too tight on the sides, and to have a fine polish.

Another point for consideration is the proper form of the journals of shafts, and that is, they should never have the journal turned or cut square down to the diameter, but hollowed in the form shown in the figure at *a a a a*. From a series of interesting experiments it has been shown that the square-cut shaft loses nearly one-fifth of its strength, and by simply curving out the shaft at the collars in the form described, the resistance to strain is increased one-fifth or in that proportion.

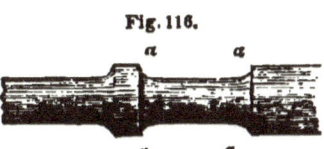

Fig. 116.

6. *Friction.*

On the subject of friction much cannot be said. We may, however, adduce a few experiments from Morin and Rivière, which appear to bear out our previous experience of the length of journals.

In the years 1831, 1832 and 1833, a very extensive set of experiments were made at Metz by M. Morin, under the sanction of the French government, to determine, as nearly as possible, the laws of friction, and by which the following were fully established:—

When no unguent is interposed, the friction of any two surfaces, whether of quiescence or of motion, is directly proportional to the force with which they are pressed perpendicularly together; so that for any two given surfaces of contact there is a constant ratio of the friction to the perpendicular pressure of the one surface upon the other. Whilst this ratio is thus the same for the same surfaces of contact, it is different for different surfaces of contact. The particular value of it in respect to any two given surfaces of contact, is called the coefficient of friction in respect to those surfaces.

When no unguent is interposed, the amount of the friction is, in every case, wholly independent of the extent of the surfaces of contact; so that the force with which two surfaces are pressed together, being the same, their friction is the same, whatever be the extent of their surfaces of contact.

That the friction of motion is wholly independent of the velocity of the motion.

That where unguents are interposed, the coefficient of friction depends upon the nature of the unguent, and upon the greater or less abundance of the supply. In respect to the supply of the unguent, there are two extreme cases,—that in which the surfaces of contact are but slightly rubbed with the unctuous matter, as, for instance, with an oiled or greasy cloth, and that in which a continuous stratum of unguent remains continually interposed between the moving surfaces; and in this state the amount of friction is found to be dependent rather upon the nature of the unguent than upon that of the surfaces of contact. M. Morin found that with unguents (hog's lard and olive oil) interposed in a continuous stratum between surfaces of wood on metal, wood on wood, and metal on metal, when in motion, have all of them very nearly the same coefficient of friction, being in all cases included between ·07 and ·08. The coefficient for the unguent tallow is the same, except in that of metals upon metals. This unguent appears to be less suited for metallic surfaces than the others, and gives for the mean value of its coefficient under the same circumstances ·10. Hence it is evident that where the extent of the surface sustaining a given pressure is so great as to make the pressure less than that which corresponds to a state of perfect separation, this greater extent of surface tends to increase the friction by reason of that adhesiveness of the unguent, dependent upon its greater or less viscosity, whose effect is proportional to the extent of the surfaces between which it is interposed.

Mr. G. Rennie found, from a mean of experiments with diffcrent unguents on axles in motion, and under different pressures, that with the unguent tallow, under a pressure of from 1 to 5 cwt., the friction did not exceed $\frac{1}{33}$th of the whole pressure; when soft soap was applied it became $\frac{1}{34}$th; and with the softer unguents applied, such as oil, hog's lard, etc., the ratio of the friction to the pressure increased; but with the harder unguents, as soft soap, tallow, and anti-attrition composition, the friction considerably diminished; consequently to secure effective lubrication, the nature of the unguent must be accommodated to the pressure or weight tending to force the surfaces together.

TABLE OF COEFFICIENTS OF FRICTION UNDER PRESSURES INCREASED CONTINUALLY UP TO LIMITS OF ABRASION. BY MR. G. RENNIE.

Pressures per square inch.	Coefficients of Friction.			
	Wrought-iron upon wrought-iron.	Wrought-iron upon cast-iron.	Steel upon cast-iron.	Brass upon cast-iron.
32·5 lbs.	·140	·174	·166	·157
1·66 cwts.	·250	·275	·300	·225
2·00 "	·271	·292	·333	·219
2·33 "	·285	·321	·340	·214
2·66 "	·297	·329	·344	·211
3·00 "	·312	·333	·347	·215
3·33 "	·350	·351	·351	·206
3·66 "	·376	·353	·353	·205
4·00 "	·395	·365	·354	·208
4·33 "	·403	·366	·356	·221
4·66 "	·409	·366	·357	·223
5·00 "	—	·367	·358	·233
5·33 "	—	·367	·359	·234
5·66 "	—	·367	·367	·235
6·00 "	—	·376	·403	·233
6·33 "	—	·434	—	·234
6·66 "	—	—	—	·235
7·00 "	—	—	—	·232
7·33 "	—	—	—	·273

From a paper lately read at the Institution of Civil Engineers in London, on the comparative friction of steam engines of different modifications, it appears that, as respects the friction caused by the strain, if the beam engine be taken as the standard of comparison—

The vibrating engine..........has a gain of 1·1 per cent.
The direct engine with slides.... " loss of 1·8 "
  Ditto with rollers........... " gain of 0·8 "
  Ditto with a parallel motion " gain of 1·3 "

It also states, as an opinion, that excessive allowance for friction has hitherto been made in calculating the effective power of engines in general; as it is found practically by experiments with the engines at the Blackwall Railway, and also with other engines, that where the pressure upon the piston is about 12 lbs. per square inch, the friction does not amount to more than $1\frac{1}{2}$ lbs.; and also that by experiments with an indicator on an engine of 50 horse-power, at Truman, Hanbury and Co.'s brewery, the whole amount of friction did not exceed 5 horse-power, or $\frac{1}{10}$th of the whole power of the engine.

7. *Lubrication.*

On this question it is necessary to observe that the durability of shafts, and their easy working, depends on the way in which they are lubricated, and the description of unguent used for that purpose. We have already seen the difference which exists in the coefficient of friction from the use of different kinds of unguents, and we have now to consider what system of lubrication should be adopted to lessen the friction and maintain smooth surfaces on the journals of shafts. In large cotton mills I have known as much as ten to fifteen horses' power absorbed by a change in the quality of the oil used for lubrication; and in cold weather, or when the temperature of the mill is much reduced (as is generally the case when standing over Sunday), the power required on a Monday morning is invariably greater than at any other time during the week.

It is, therefore, necessary in most mills—particularly those employed in textile manufacture—to retain a uniform temperature, and to employ the best quality of oil for lubricating the machinery, as well as the shafts of the mill.

The best lubricators are pure sperm and olive oils; they should be clean and limpid, and sparingly applied, as it is a profligate waste of valuable material to pour, as is not unfrequently done, large quantities of oil on the bearings, nine-tenths of which run on to the floor, and cover the shafts and hangers with a coat of glutinous matter, that soon hardens, and accumulates nothing but filth.

This process of oiling shafts is generally left to the most negligent and most untidy person in the establishment; and the result is, that every open-

STRENGTH AND PROPORTIONS OF SHAFTS. 215

ing for the oil to get to the bearings is plugged up, the brass steps are cut by abrasion, and the necks or journals of the shafts destroyed. In the best regulated establishments this is certainly not the case, as the greatest possible care is observed in selecting the best kinds of oil, and that used with attention to cleanliness and strict economy in its application.

To save power and effect economy in the use of lubricants, several schemes have been adopted for attaining a continuous system of lubrication. None of them appears to answer so well as that which consists of a small cistern, a, fig. 117, which contains a quantity of oil, and is fixed on the top of the plummer block. In the centre of the cistern is a tube, which stands a little above the level of the oil; and into this is inserted a woollen thread, with its end descending a short distance below the surface of the oil in the cistern; and when properly saturated, the oil rises by capillary attraction, and flows gently, in very minute quantities, on to the neck of the shaft. From this description it will be seen that the quantity used can be regulated to the greatest nicety, and sufficient to lubricate the bearings without waste. Other plans have been devised for the same object, but none of them seems to answer so well as that just described.

Fig. 117.

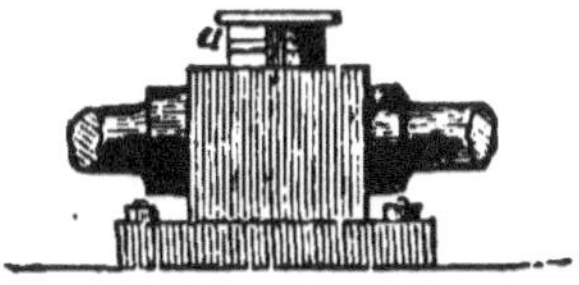

CHAPTER IV.

ON COUPLINGS FOR SHAFTS AND ENGAGING AND DISENGAGING GEAR.

In every description of mill where the machinery is spread over a large area, and at a distance from the moving power, it is necessary to have long lines of shafting, revolving at the required velocity. Such lines are seldom made in one piece; short lengths must, therefore, be coupled together, so as to form an unbroken line, extending, in most cases, the whole length of the mill.

When cast iron shafts were substituted for wood, a square coupling-box, made in one piece, was generally used, so as to slide over the two ends of the shafts, or in two pieces, bolted together, as shown in figs. 118 and 119.

Fig. 118.

In the former case the box was slipped on loose, and the adjustment was so imperfect that the shafts rose and fell in the box at every revolution, destroying gradually any accuracy of fitting which, in the first instance, had been attained.

ON COUPLINGS.

After the square-box coupling came the claw, or two-pronged coupling, made in two parts,

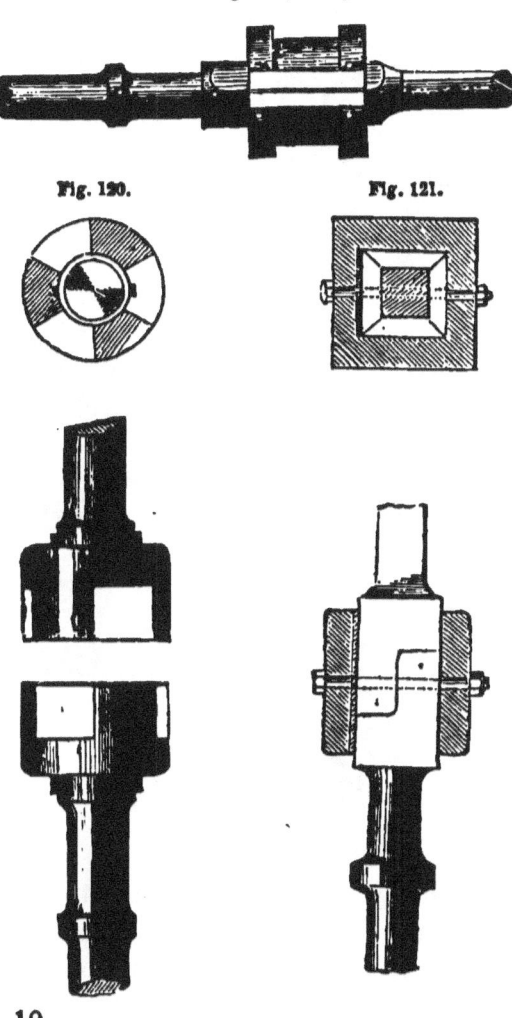

Fig. 119.

Fig. 120.   Fig. 121.

wedged, but more frequently keyed on to the ends of the shafts, as shown in fig. 120. This was a great improvement, as the leverage of the bear-

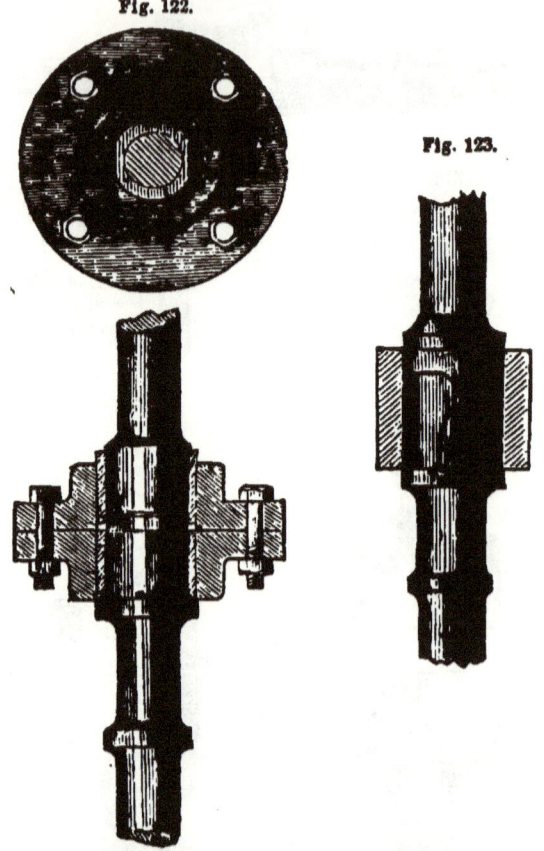

Fig. 122.

Fig. 123.

ing parts was greatly increased, and the coupling, in consequence, became more durable.

A description of half-lap coupling was introduced by the late Mr. Hewes. It was formed by the lapping over a part of the end of each shaft, which was cast square. A square box was also fitted over the two ends, so as to bind them together, and *three* keys were inserted on the top side, as shown in fig. 121. The objections to this coupling were the difficulty of fitting and the loosening of the keys, which made a creaking noise with every revolution of the shaft.

Another coupling, still in use, is the disc. It consists of two discs or flanches, one on the end of each shaft, bolted together by four bolts, as shown in fig. 122. This coupling was superior to all the preceding, when properly bored and turned, so as to have its faces accurately perpendicular to the shafting.

The best coupling for general purposes, and the most accurate and durable, is the circular half-lap coupling, introduced into my own works nearly forty years ago. It is perfectly round, and consists of two laps, turned to a gauge, and, when put together by a cutting machine, it forms a complete cylinder, as shown in fig. 123. A cylindrical box is fitted over these, and fixed by a key, grooved half into the box and half into the shaft. The whole is then turned in the lathe to the same centres as the bearings of the shaft, and by this process a degree of accuracy is attained which cannot be surpassed, nor is any other coupling so neat and so well adapted for the transmission of power.

The proportions of this coupling are found by experiment to be—

Twice the area of the shaft is the area of the coupling.

The length of the lap is the diameter of the shaft.

And the length of the box is twice the diameter of the shaft.

These proportions have been found in practice to answer every purpose, both as regards strength and the wear and tear of the joints.

There is another coupling which has come of late years extensively into use, namely, the cylin-

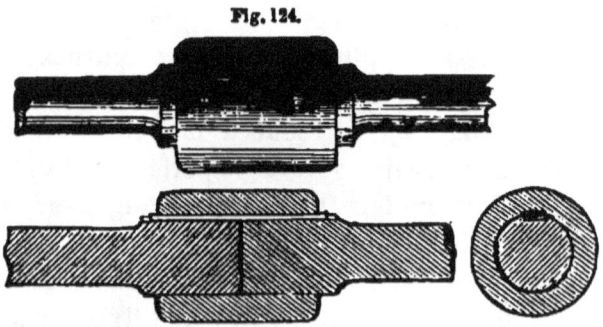

Fig. 124.

drical coupling, with butt ends. It has the same proportions as the former, but not so strong nor so durable as the half-lap coupling of the same dimensions, as the entire force of torsion is transmitted through the key; but in cases where strength is not the chief object, it forms a cheap and effective coupling.

8. *Disengaging and re-engaging Gear.*

This is an important branch of mill-work, requiring careful consideration, and the utmost exactitude of construction when ponderous machinery has to be started, without endangering the shafts and wheels. This is most strikingly exemplified in the case of powder mills, where trains of edge stones are employed for grinding the gunpowder, and in rolling and callendering machinery which requires well fitted friction-clutches to communicate the motion by a slow and progressive acceleration from a state of rest to the required velocity.

It used to be customary in cotton and silk mills to place disengaging clutches at the point of connection of the upright or driving shaft and the main shafting in each room, so that, in case of accident, a room full of machinery could be thrown out of gear at once. But these provisions were found unsteady in practice, and rather tended to increase than to diminish the number of accidents, owing chiefly to the time lost in disengaging, and the breakages which occurred in attempting to place the machinery in gear again, when the engine was running at full speed. It has, consequently, been found safer to have a permanent connection between the main lines of shafting throughout the mill, and signals from each room into the engine-house, in case of accident.

When the construction of mill gearing was less

222   MACHINERY OF TRANSMISSION.

perfect than it is at present, the main shaft driving the machinery in a room was thrown out of gear by a lever, which contained the steps, and supported the end of the horizontal shaft and wheel,

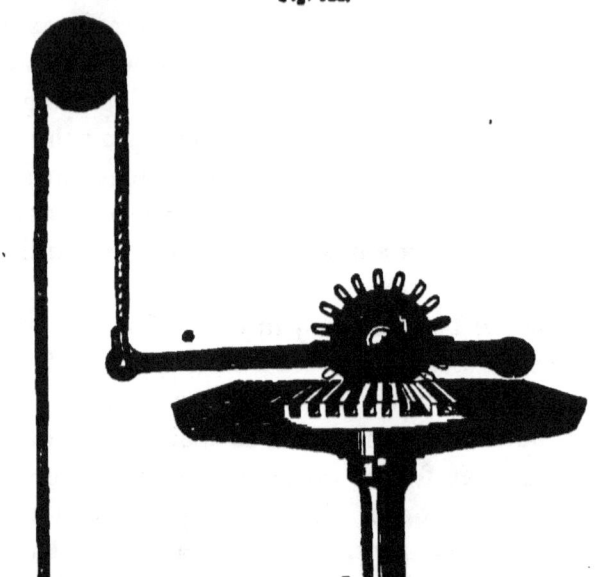

Fig. 125.

which geared into that on the upright shaft, as shown in fig. 125, with a rope at the end of the lever *a* to pull it out of gear. This mode of disengaging wheels was very ineffective, as in many mills there are three bevel wheels gearing into that on the upright *b*, and it becomes complicated and dangerous to have movable levers to each. To remedy these defects, standards or plummer-blocks, with a movable slide e, fig. 126, in which the end of the shaft revolved, was introduced. To the top of this slide was attached a lever a, with a handle b, by which it could be drawn out of gear; and the link c, falling along with the lever, re-

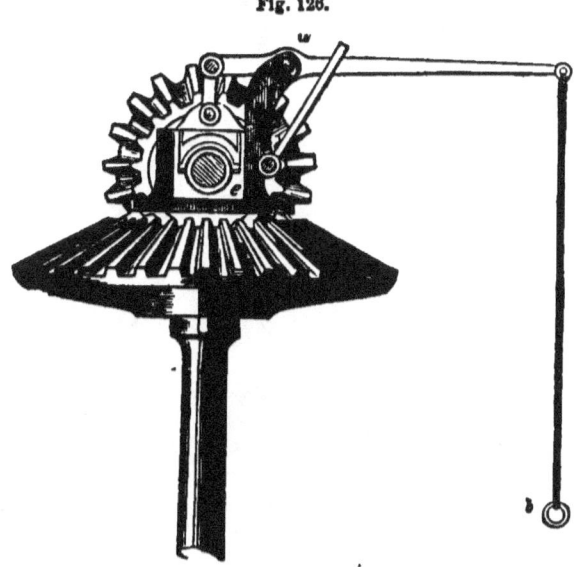

Fig. 126.

tained the shaft out of gear until the mill was stopped.

All these contrivances were, however, found inoperative on a large scale, as the shafts and wheels got out of order; and it was ultimately found essentially necessary to make them stationary, by screwing the plummer-block down to the frame which connects the shafts and wheels.

Several devices have been employed for the purpose of rapidly engaging and disengaging machines from the driving shaft. The best of all are

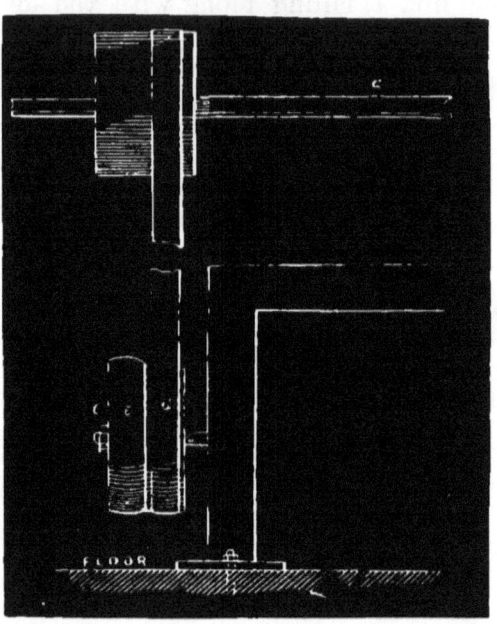

Fig. 127.

the fast and loose pulleys, with a travelling strap. Thus, in fig. 127 *a* is the driving shaft, acting upon two pulleys *e* and *d*, fixed on the driving spindle of the machine *b*; one of them, *d*, is keyed fast, and the other runs loose. When the machine is at work the strap is on the fast pulley *d*, and when it is necessary to stop, it is moved by a forked lever on to the loose pulley e, which revolves with the strap without acting on the machine. The machine is thrown into gear with equal ease by moving the strap on to the fast pulley d. Once on either of the pulleys, the strap is held in position without any danger of moving by the slight curvature of the pulley, as already explained. The forked lever must act on that side of the strap which runs toward the pulleys, and not on that which leaves them.

A second and equally effective process for start-

Fig. 128.

ing or stopping machinery is shown in fig. 128. A leather strap is hung loosely over the driving and driven pulleys *a* and *b*, so that, left to itself, the friction is not sufficient to communicate motion to the pulley on the shaft *b*; but a tightening pulley fixed on a suitable lever *e* is forced against it by pulling the rope *c*, which bends the strap tightly upon the pulley *b*, and gives motion to the machine. This arrangement is in general use for sack teagles in corn mills, and for some other purposes. The same effect is sometimes produced by the sack teagles being fixed on the lever, and, by raising one end, the strap is tightened, and the barrel which raises the load is caused to revolve.

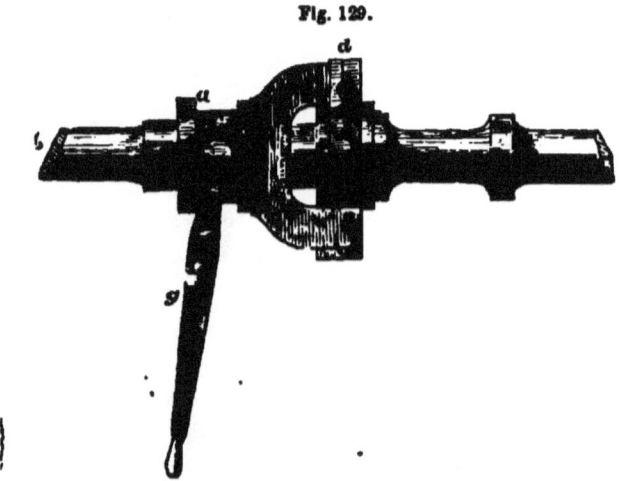

Fig. 129.

The clutch most in use for throwing into gear heavy callendering machines is a clip friction hoop, which consists of a sliding box a, with two projecting horns on the driving shaft b. These horns, when slid forward by a lever g, working in the groove c, come in contact with the friction hoop d, which embraces a groove in a second box, keyed upon the shaft of the machine. The instant the machine receives the shock of engagement, the clip d slides in its groove, until the friction overcomes the resistance, and the callender attains the speed of the driving shaft. The object of the friction clip is to reduce the shock of throwing the clutch suddenly into gear, as without this precaution any attempt to move instantaneously a powerful machine from a state of rest to a state of motion would break it in pieces.

Friction cones are also much used for this purpose, and when carefully executed with the proper angle are safer than the clutch just described. The objection to the friction clutch is, that the whole driving power is thrown on the clip at once; whereas, with the cones, the parts can be brought into contact with the greatest nicety, and the friction regularly increased to any degree of pressure. Fig. 180 shows this description of disengaging gear; a is the male sliding cone, worked by a lever in the usual way, b the female cone, keyed on the driven shaft, and the two surfaces, when brought into contact, communicate the required motion with perfect safely.

Machines driven by friction, and requiring to

228 MACHINERY OF TRANSMISSION.

Fig. 130.

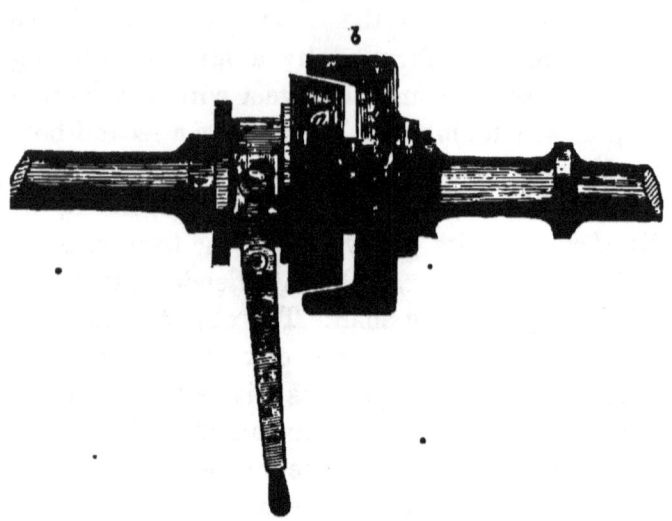

be frequently stopped, are very numerous. Some of the lighter description are driven by a vertical shaft, *b*, fig. 131, supporting a horizontal disc,

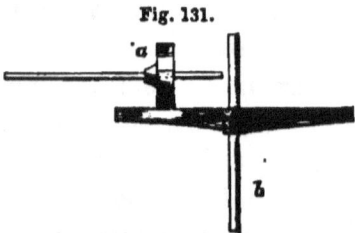

Fig. 131.

which communicates motion to the wheel *a*, rolling on its surface, and gives the necessary motion to the machine. The advantage of this friction-wheel is, that the velocity of the machine may be increased or diminished at pleasure by moving the wheel *a* nearer to or farther from the edge of the disc.

ENGAGING AND DISENGAGING GEAR. 229

Fig. 132 is another combination of discs suitable for couplings with only one bearing. The disc *b* is keyed on one shaft, and is recessed on the face, to receive the smaller disc, *c*; this disc is sunk flush with the face of the other, and is screwed tightly up to it by means of the ring *a*,

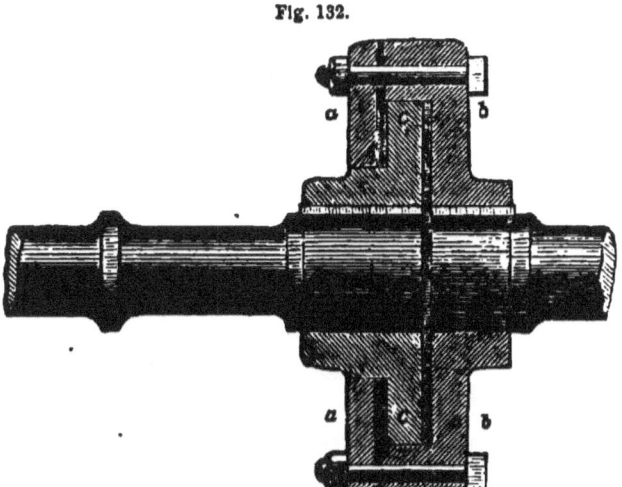

Fig. 132.

which is bolted to the disc *b*, and secures that marked *c*. Between the three plates, *a*, *b* and *c*, annular pieces of leather are interposed, which bring them all to a proper bearing.

This combination, termed a friction coupling, is useful for preventing breakage of the connections in case of a sudden stoppage or reversal of the motion. It is plain that the holding power of the coupling depends simply upon the lightness with which the discs are screwed together, and the consequent frictional force of the surfaces of leather and metal.

Besides these more permanent forms of couplings, there are other contrivances adopted when the object to be attained is the engagement and disengagement of certain parts of the machinery or gearing during the course of operations.

With the same view of admitting of this disengagement of the connection, in cases of sudden

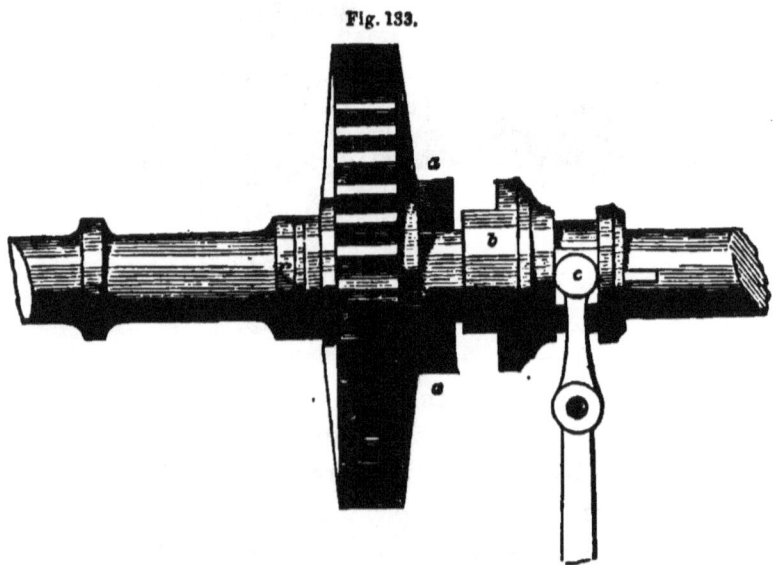

Fig. 133.

stoppage or reversal, the coupling, fig. 133, is sometimes employed.

In this instance, the shaft is supposed to be continuous, and the coupling may be termed a disengaging coupling. *a* and *b* are the two parts of the coupling, formed on the acting faces into alternate projections and recesses, such that they correspond with and exactly fit into each other when in gear. The part *a* is, in this example, cast on a spur or bevel wheel, from which the motion of the shaft is supposed to be taken off. Both of the parts *a* and *b* are, to a certain extent, loose on the shaft; the former being capable of moving round on it, though deprived of longitudinal motion by washers and a collar, marked *e*, and the latter being free to slide on the shaft, though prevented from turning on it by a sunk key, which slides in a slot inside the clutch or sliding piece *b*. The mechanism is put into gear by means of the lever *d*, which terminates in a fork with cylindrical extremities *c*; and it is obvious that, by the contact of the flat faces of *a* and *b*, the latter will immediately carry with it the other part at the same speed as the shaft. Supposing, now, that the motion of the wheel *a* is suddenly accelerated, the oblique faces of the couplings immediately fall out of contact, and slide free of each other, leaving the couplings clear, and the shaft free to continue in motion.

In the old form of this contrivance, known as the sliding bayonet clutch, the part *b*, instead of the toothlike projections on the face, had two or more prongs which laid hold of corresponding snugs cast on the face of the part *a*, which, more over, was usually a broad belt pulley, introduced with a view to modify the shock on the gearing on throwing the clutch into action.

In an older form still, the pulley was made to slide end long on the shaft. A form analogous to this was known as the "lock pulley," a few specimens of which still remain in the older factories. Instead of the end long motion common to the other modes, the parts were "locked" together by a bolt fixed upon the side of the pulley, and which, when shifted toward the axis, engaged with an arm of a cross, of which the part b, in the preceding figure, is the modern representative. The bolt was wrought by means of a key and stop, the turning of the key throwing back the bolt, and thereby unlocking and disengaging the pulley. The form of coupling represented by fig. 133 is particularly applicable when the impelling power is derived from two sources—a circumstance which frequently occurs in localities affording water power to some extent, and yet not in sufficient abundance for the demands of the work. The deficiency is usually supplied by a steam-engine; and the two powers are concentrated in the main line of shafting by a coupling of the kind depicted. In cases of this kind, the speed of the shafting being fixed, and the supply of water inconstant, the power of the water-wheel ought to be thrown upon the wheel $a\ a$, and that of the engine upon the shaft at another point. By this arrangement, the speed of the line can be exactly regulated by working the engine to a greater or less power, according to the supply of water. The proper speed of the water-wheel will likewise be maintained, which is of importance in economising the water power.

"The same form of coupling is also used occasionally for engaging and disengaging portions of the machinery. But for this purpose the object is to obtain a mode of connection by which the motion may be commenced without shock; for, in consequence of the inertia of all material things—that is, the tendency which every portion of matter has, when at rest, to remain at rest, and when in motion, to continue to move—the parts of the mechanism, when acted upon too suddenly by a moving power, are liable to fracture and disarrangement. It is a law in mechanics that when a body is struck by another in motion some time elapses before it is diffused from the point struck through the other parts; consequently, if the parts receiving the blow have not sufficient elasticity and cohesive force to absorb the whole momentum of the striking body till the motion be transmitted to the centre of rotation, fracture of the body struck must necessarily ensue. Hence, in a system of mechanism, any parts intended to be acted upon suddenly by others in full motion ought not only to be strong, but they ought to be capable of yielding on the first impulse of the impelling force with as little resistance as possible, and gradually bring the whole weight into motion. The common mode of driving by belts and pulleys accomplishes this object very satisfactorily. In this the elasticity of the belt comes into action; and being thrown upon the pulley by the strap guide or fork, it continues to slip, till, by the friction between the sliding surfaces, the belt gradually brings the quiescent pulley into full motion. This mode of connection is unexceptional when the power to be transferred is not great; but its application to large machinery is attended with inconvenience." *

In figs. 134 135, two other forms of clutches are shown, as often used to connect the shafting of different parts of the same mill, where it is not necessary to throw into or out of gear when running at full speed. They consist of a fixed and sliding box, one on each shaft, with teeth or projections which fit in corresponding notches. The sliding box has a groove turned in it, in which a forked lever works, as at *a*, fig. 134, and at *a*, fig. 135, by which it is drawn backward or forward as the case may be. The peculiarity of the clutch, fig. 135, is that of the driving shaft, which, reversed by any accident in its motion, as is not unfrequently the case in starting and stopping the steam engine, the sliding clutch is forced back

---

* Extract from Engineer's and Machinist's Assistant, p. 144.

by the wedge-shaped faces of the projections, and the machinery thrown out of gear.

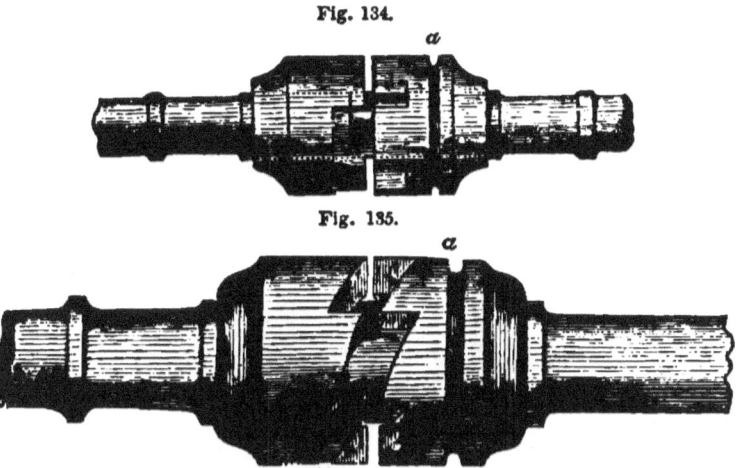

Fig. 134.

Fig. 135.

Fig. 136 shows one of these clutches on a small scale, fixed on a line of shafting beneath the floor of a mill. It is placed between two standards *a a*, supporting the ends of the shaft, and the lever *b* working on a pivot at bottom, and having a pin working in the groove of the sliding clutch box, serves

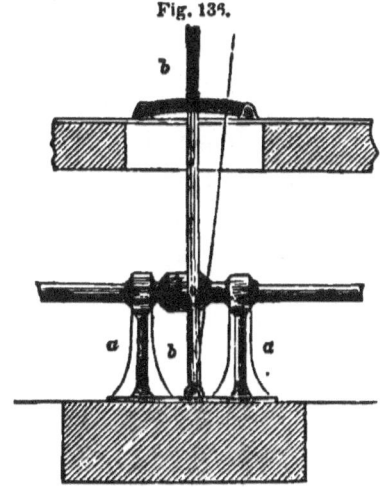

Fig. 136.

for throwing the driven shaft into or out of gear whenever it may be necessary.

Another ingenious contrivance, I believe invented by Mr. Bodmer, is shown in figs. 137 and 138. It consists of a box *a a* running loosely on the driving shaft *s s*, but carrying the bevel wheel *b b*, which gears into another wheel on the driven shaft, not shown in the figures. Tightly keyed

Fig. 137.

on the driving shaft *s s* is a boss *c c*, with two trunnions, on which slide two friction sectors *k k*; the outer surface is coated with a copper plate, accurately fitting the interior surface of the running box *a a*. The boss *c c* carries also four projections *e e e e*, which serve as guides for four screws, alternately left and right handed, and attached to the nuts *f f* and levers *g g*; these screws act on the extremities of the friction slides *k k*, so that when the levers *g g* are drawn back they are both with equal pressure forced upon the inner

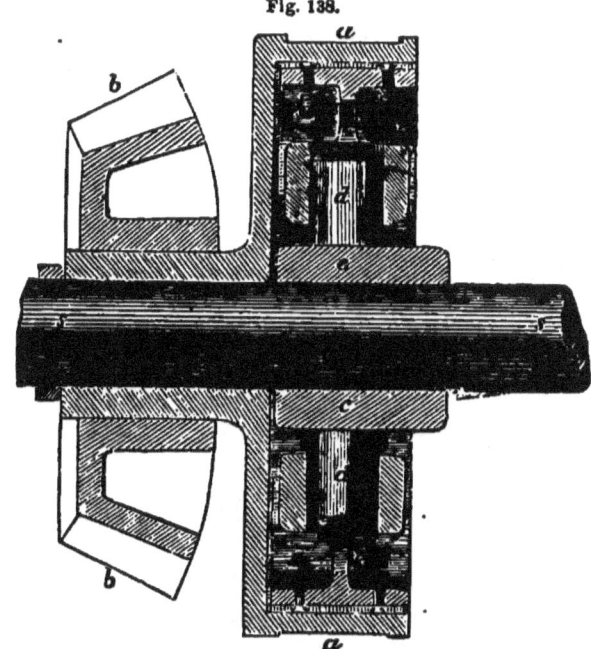

Fig. 138.

surface of the box *a a*. As the pressure can be very regularly and gradually brought on this box through the levers and screws, the motion of the driving shaft *s s* is communicated with perfect regularity, and without shock to the bevel wheel *b b*.

In the above description I have given such examples of engaging and disengaging gear as are most commonly in use. Others of a more complicated character might be cited, but they are not to be recommended as applicable in general practice. The last form, figs. 137 and 138, is, however, specially noticed as suitable for gunpowder mills, where the greatest possible freedom from shocks is essentially necessary.

9. *Hangers, Plummer-blocks, etc., for carrying Shafting.*

Shafting is supported in three ways, viz., on foundation stones in the floor, beneath beams suspended from the ceiling, and to the walls of the mill. This necessitates as many different forms of framework, known as hangers, plummer-blocks, standards, etc.

The simplest mode of supporting a range of light shafting is from the floor, and a pedestal suitable for this purpose is shown in fig. 139. It consists of a cast iron base plate and column, with deep wings *a a* cast on to strengthen it free from vibration. The upper portion is hollowed out to receive the lower brass step, and the cap carrying

SHAFT HANGERS AND PLUMMER-BLOCKS. 239

the upper step. When the entire pressure of the shafting is downwards the upper brass bush is omitted, and the cap is cast hollow and kept full of grease, so as to secure the most perfect lubrication of the journal of the shaft.

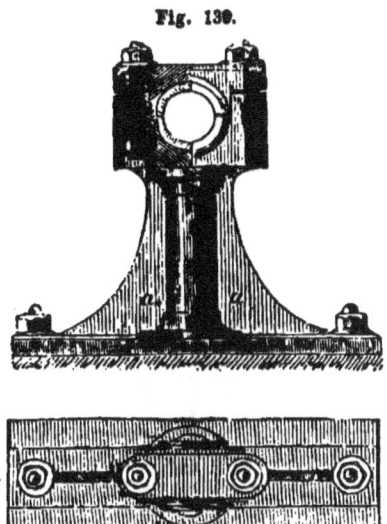

Fig. 139.

Fig. 140 shows a pedestal for bolting to a wall, the chief difference being that the cap is now fixed on its inner side by a wedge or cotter (c). In this figure a shell cap *a* is shown. If the pull is upwards, and two brasses be required, "lugs" have to be added to the extremity of the pedestal and cap for bolting the two together.

There are various ways of suspending ranges

240   MACHINERY OF TRANSMISSION.

of shafting from the ceiling, according to the means which exist for their attachment. If wooden beams, as s, are present, the hanger has a large

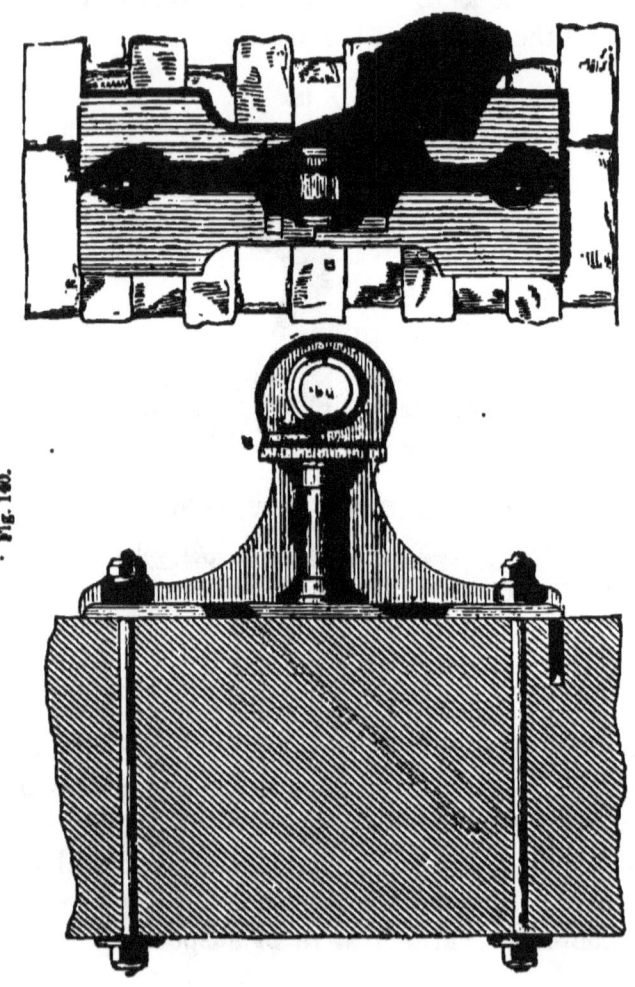

Fig. 140.

SHAFT HANGERS AND PLUMMER-BLOCKS. 241

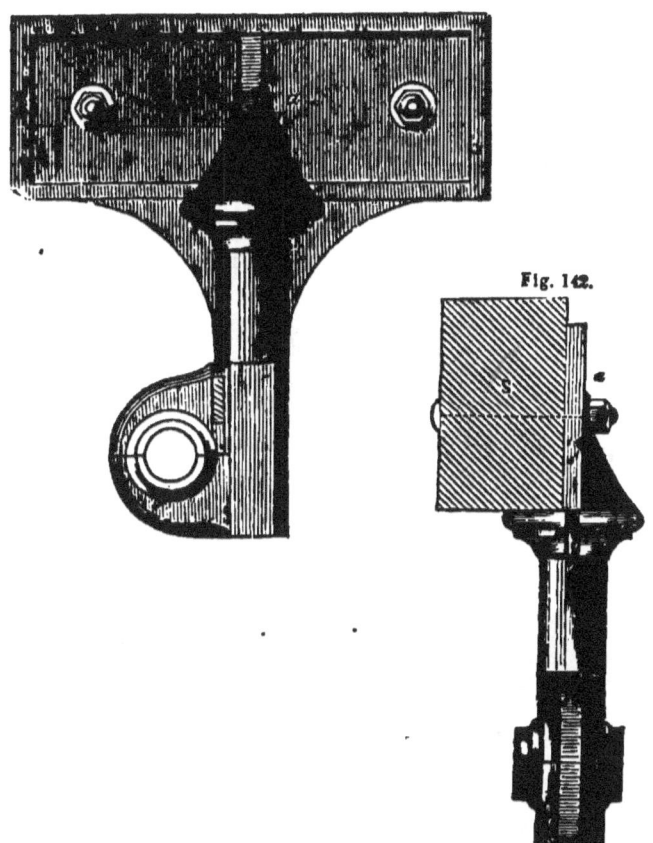

Fig. 141.

Fig. 142.

plate (*a*), which bolts to the side of the beam, as shown in figs. 141, 142. The caps are fixed by a cotter, as in the previous case.

Figs. 143, 144, show a front and side elevation of another form of hanger for attachment to wooden

Fig. 143.

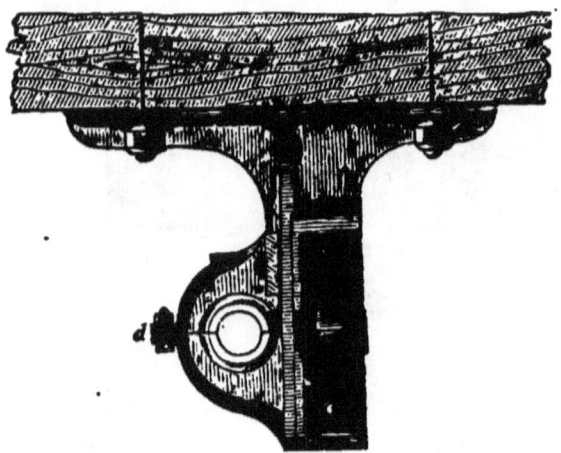

beams In this case there is provision for a second line of shafting, at right angles to, and receiving motion from, the primary line. For this purpose a small plummer-block is bolted on to a recess at the side of the hanger. The thrust, owing to the pair of bevel wheels which would be placed near this hanger, is no longer simply vertical, and hence two brass steps are placed for the journal of the principal shaft, with a bolt at d, fig. 143, in addition to the cotter, to keep the cap in its place.

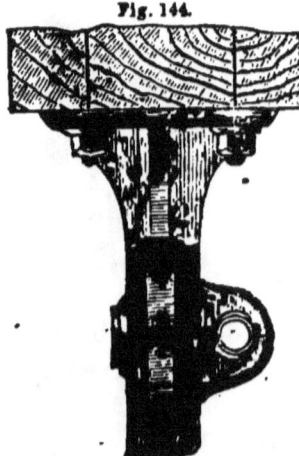

Fig. 144.

SHAFT HANGERS AND PLUMMER-BLOCKS. 243

Fig. 145 shows another form of light hanger sometimes employed in weaving sheds, and also in use for supporting shafts in fire-proof mills,

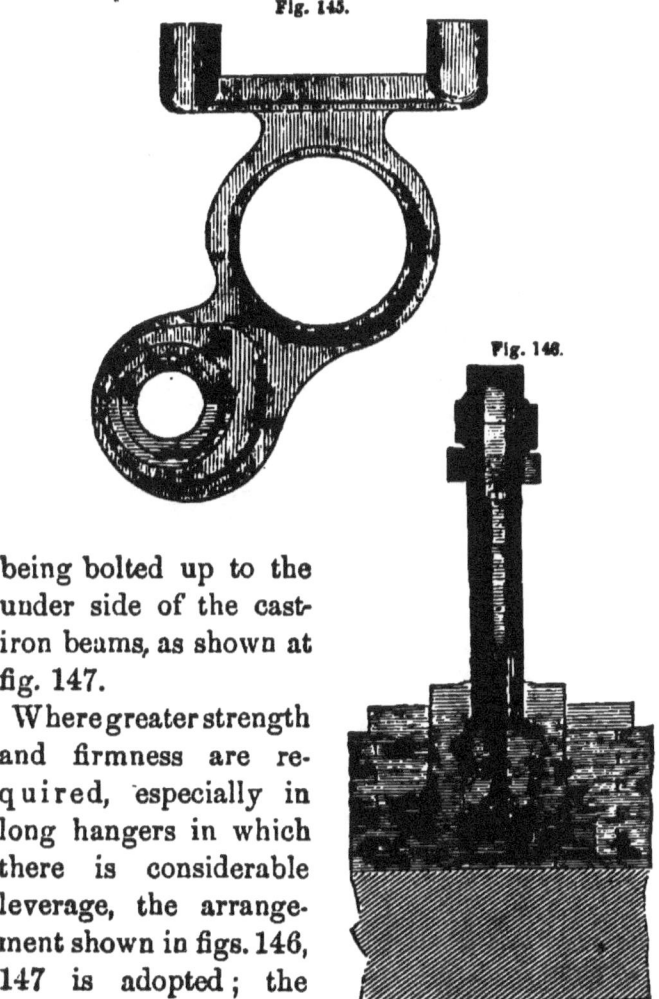

Fig. 145.

Fig. 146.

being bolted up to the under side of the cast-iron beams, as shown at fig. 147.

Where greater strength and firmness are required, especially in long hangers in which there is considerable leverage, the arrangement shown in figs. 146, 147 is adopted; the hanger in this case is bolted to a cast-iron beam, and by an extension of the flange plate to the brick arch, which springs from the beam T, it is firmly secured to both beam and floor. At *e* is a screw for tightening the upper brass step on the shaft.

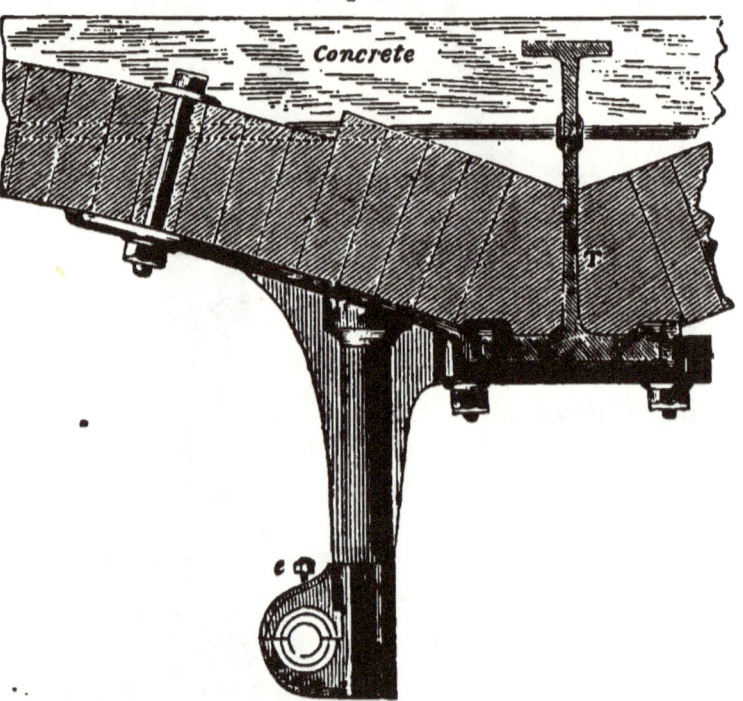

Fig. 147.

More complicated arrangements are sometimes necessary where two or three ranges of shafting have to be brought in connection with each other

SHAFT HANGERS AND PLUMMER-BLOCKS. 245

by means of bevel or mitre wheels. Figs. 148 and 149, show a front and side elevation of this arrangement, which may serve as a type for others. The hanger is attached to a cast-iron

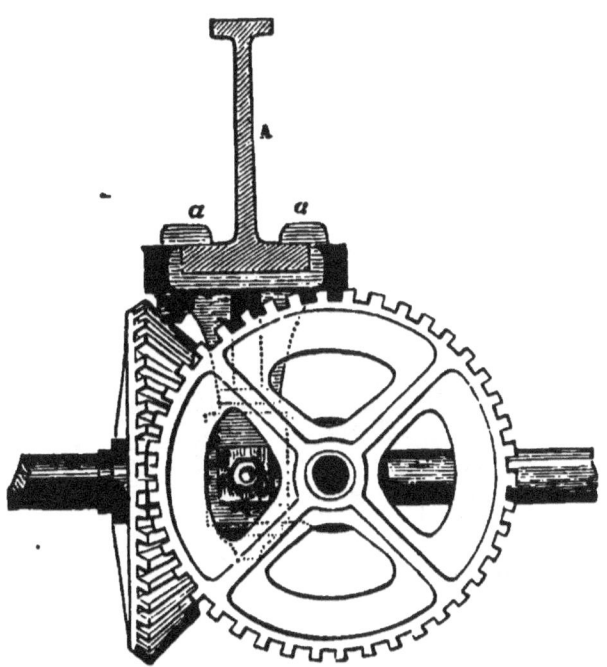

Fig. 148.

beam A, by hooked bolts with nuts beneath the top plate, as shown at *a a, care being taken in this attachment not to weaken the flange of the iron beam by boring holes in it.* Double brass steps are necessary in this case for the main line of shafting, and also for two smaller ranges at right angles to it,

21*

which revolve in opposite directions, as shown at fig. 149.

A very frequent case in practice is the connection of two ranges of shafting, at right angles to

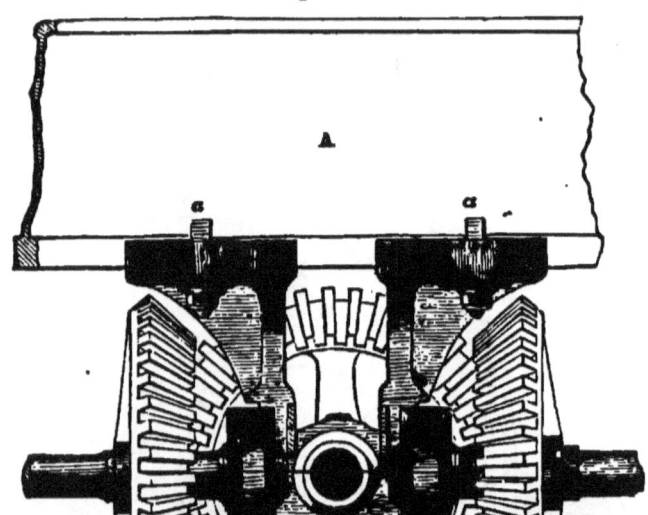

Fig. 149.

each other, at the corner of a room. This is effected by letting into the corner of the building a cast-iron frame, commonly known as a wall-box, which serves as a foundation for the plummer-blocks carrying the shafting. Such an arrangement is shown in fig. 150 in elevation, and in fig.

151 in plan. The box w, w, w, is built into the wall, and bolted both to it and to the cast-iron beam b. It carries two plummer-blocks on a plate firmly supported by brackets. The wall

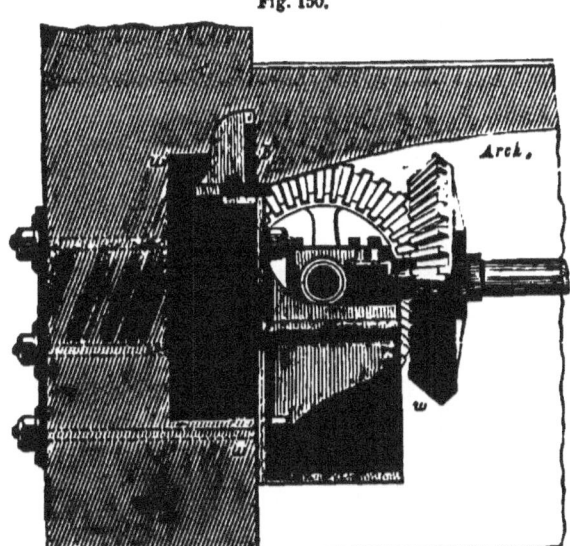

Fig. 150.

pieces in these two figures are similar, but with a slightly different arrangement of the plummer-blocks.

Irrespective of the various forms of engaging and disengaging apparatus, it will be necessary to consider the position, form, and proportions of the wheels and shafting required in mills where the power is divided and widely distributed. To show the enormous extent to which the concentration of machinery in one building has been carried, I may mention that in mills of my own construction there have been on the average not less than 450 wheels and 7,000 feet of shafting in motion. In the large mills at Saltaire there are upwards of 600 wheels and 10,000 feet, or two miles, of shafting distributed over an area of

Fig. 151.

flooring equivalent to 12 acres. In corn mills and iron works, where the machinery is more closely connected with the prime mover, these considerations are of less importance; but in factories for the manufacture of textile fabrics the machinery covers a great extent of surface, and the greatest care is necessary in giving due proportion to the transmissive machinery, in order to secure uniformity of motion at the remotest parts of the mill.

In gearing a mill, the first consideration is the power of the engines, the position of the machinery to be driven, and the strength, diameter, etc., of the first-motion shaft, and other requisites for the transmission of motion in a well-geared mill. It is upwards of twenty years since the fly-wheel was converted into a first motion, and a new system of transmitting the power of the steam engines to the machinery of the mill introduced. Previous to that time it was effected by large spur-wheels inside the mill, now it is taken direct from the circumference of the fly-wheel.* The advantage of this system was the abolishing of the cumbrous first-motion gearing; and the requisite velocity being already present in the fly-wheel, it was only necessary to cast it with teeth, and to take off the power by a suitable pinion at the level most convenient for the purposes of the mill.

In another place I have given general rules for the pitch, breadth, and strength of the teeth of wheels. The Table (p. 250), computed from examples which have occurred in my own practice,

---

* Compare "Mills and Millwork," Part 1, Prime Movers, page 248.

exhibits the best proportions of spur fly-wheels to secure durability of both wheel and pinion.

TABLE 9.—DIAMETERS, PITCH, VELOCITY, ETC., OF SPUR FLY-WHEELS OF THE NEW CONSTRUCTION.

Nominal power of Steam Engine.	Diameter of fly-wheel.	Pitch in inches.	Breadth of cog in inches.	Velocity of pitch line per minute in feet.
Horse-power.	Ft. Ins.			
Two 150 = 300	30 1½	4½	16	The velocities vary according to circumstances, from 1,250 to 1,650 feet per minute.
Single = 50	13 3¾	4⅜	12	
Two 100 = 200	24 5	4	14	
Two 80 = 160	23 4	4	14	
Two 80 = 160	22 4	4	14	
Single 60	19 0½	3⅜	12	
Two 70 = 140	24 5	3¼	12	
Two 70 = 140	22 8¼	3¼	14	
Two 50 = 100	21 0	3¼	12	
Two 40 = 80	21 0	3¼	10	
Two 45 = 90	20 0	3¼	12	
Single 50	18 2½	3¼	12	
Two 35 = 70	16 0¾	3¼	10	
Single 40	17 10	3	10	
Two 25 50	13 10	3	10	
Single 25	8 11½	3	12	
Two 20 = 40	15 6	2½	7	
Two 25 = 50	15 4½	2½	8	
Single 25	15 4½	2½	7	
Two 18 = 36	13 0	2½	8	
Single 15	10 0	2½	7	
Single 18	17 11	2	6	
Single 12	10 0	2	5	

It will be observed that the diameters of the fly-wheels are not always proportionate to the power of the engines, nor yet to their respective velocities. In practice, it is impossible to maintain uniformity in this respect, as, in order to meet all the requirements of manufacture, it is necessary to deviate from fixed principles, and to approximate as near as circumstances will admit to the diameters, weights, and velocities of wheels, as may be found convenient to produce a maximum effect.

Of late years, the speed of the piston of factory steam engines has been accelerated from 240 to 300, and in some cases to 350 feet per minute. This united to the increased pressure of steam nearly doubles the power of the engines to what they were thirty years ago. The standard speed of a Bolton and Watt 7 feet stroke engine previous to that date, was seventeen and a half strokes per minute.

In closing this section of practical construction, I may state that the couplings, engaging and disengaging gear, including the different forms of hangers, fixings, etc., are taken from my own designs, first introduced as a substitute for the cumbrous attachments that were in general use previous to the years 1820 and 1823.

Having determined the diameter, speed, and strength of the fly-wheel, the next consideration is the material, diameter, etc., of the main shaft. These are usually of cast-iron, and their diameters depending on the power transmitted through them, and the velocity at which they revolve, will be found by the tables and formulæ already given

The distribution of the power is usually effected by a vertical shaft, extending from the bottom room, through the various floors of the mill, to the top story; the power being taken off at each stage by a pair of bevel wheels. This arrangement, as shown in fig. 152, represents one engine-house with a section of part of one division of the mills at Saltaire; and this may be considered as a type of other mills adapted for spinning and similar purposes.

It will be observed that there are four divisions in the Saltaire mills—one for the preparatory process, one for the wool combing, another for the spinning, and a fourth for the weaving. All these are driven by four steam engines, each of 100 nominal horses' power,. but collectively distributing a force through these different departments of upwards of 1,250 horses.

On referring to the drawings, figs. 152 and 153, which represent a cross and longitudinal section of the mill, it will be seen that the vertical shaft AA, is driven direct from the fly-wheel by the horizontal shaft B, giving motion to the machinery in each room as it ascends. It is fixed on a solid pier of ashlar, as shown at fig. 154, page 257, and supported on strong cast-iron plates and bridgetrees, firmly secured by bolts to the foundations below. In each room it is securely fixed, by cast-iron frames and boxes,

Fig. 152.

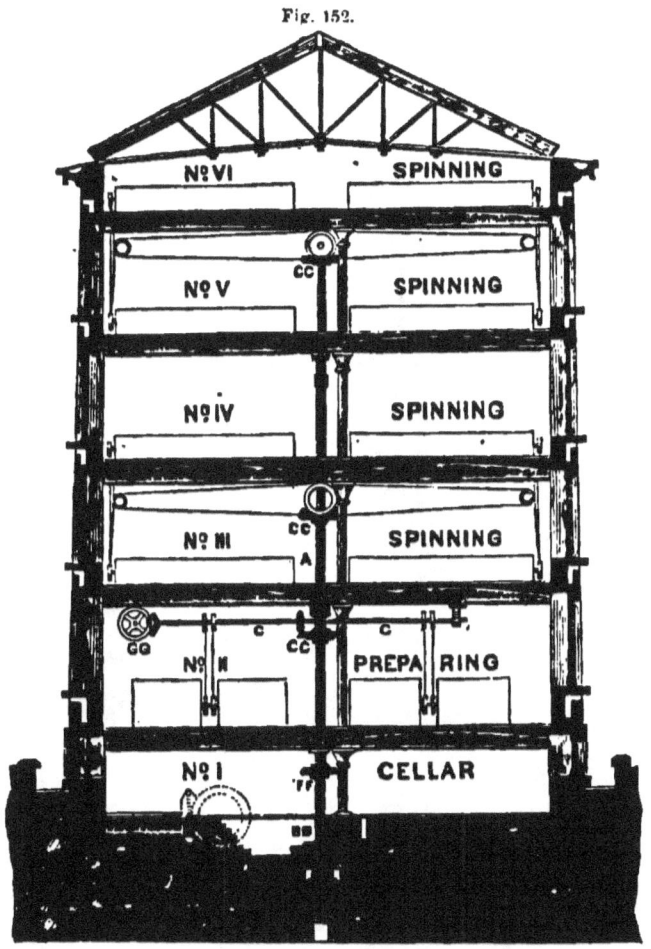

forming a recess for the bevel wheels, into the wall which divides the engine-house and the rooms above from the mill. This wall is generally made strong and thick, with sufficient

254 MACHINERY OF TRANSMISSION.

weight to resist the action of the wheels prepared to drive the main lines of horizontal shafts with a speed and force equivalent to the work done in each room. In the case of the Saltaire mills this is considerable; nearly 300 horses' power being distributed through the upright shaft alone, the remainder being carried off to the loom shed by a second wheel, working into the bevel wheel a, on the horizontal shaft B, but not shown in the drawing. It is important, in mills where powerful steam engines are employed, that the foundations and fixings to which the main shafts are attached are of the most substantial description, and the greatest precaution is necessary in order to secure them from vibration, and to render them perfectly rigid when the whole force of the engines is applied.

In the Saltaire mills, as in many others for the manufacture of cotton, flax, and wool, the preparatory machinery, such as carding, combing, roving, etc., is generally driven by lines of horizontal shafts, or by a series of cross shafts, branching off at right-angles from the main line extending down the centre of the room, as shown at $c c$ in No. II. room. Nos. III. and IV. rooms are driven by the longitudinal shaft in No. III.; and Nos. V. and VI. by the shaft in No. V. room. On this plan it will be noticed that the spinning machinery is driven by iron pulleys from the horizontal shafts, at a velocity of nearly 200 revolutions per minute, and the straps or belts from those pulleys are directed by means of guide pulleys to the machinery in both rooms. For this purpose, iron boxes are inserted into the arches supporting the floors, for the admission of the straps to the machinery in the upper floor.

It will not be necessary to give the dimensions of the shafts in each room, as these details and calculations must be left to the judgment of the millwright, and the nature of the work they have to perform. Suffice it to observe, that the vertical shaft A is 10 inches diameter through the first two rooms, 8½ inches through the third room, and 6½ inches to the top; the velocity being 94 revolutions per minute.

As respects the couplings for this shaft, we may refer the reader to the Table of Dimensions (page 109) made from couplings actually in use, and which have been found, by experiment, serviceable in every case where strength and durability are required.

Great trouble is sometimes experienced with the foot of the vertical shaft, which, from its weight and the great pressure upon it, has a tendency to heat, unless sufficient bearing-area is allowed, and the parts kept thoroughly lubricated. The general arrangement of the footsteps and gearing in large mills is shown in fig. 154; *s s* is the first-motion shaft, and *t t* the vertical shaft; *a a* the bevel wheel on the former, and *b b* the

VERTICAL SHAFTS. 257

Fig. 154.

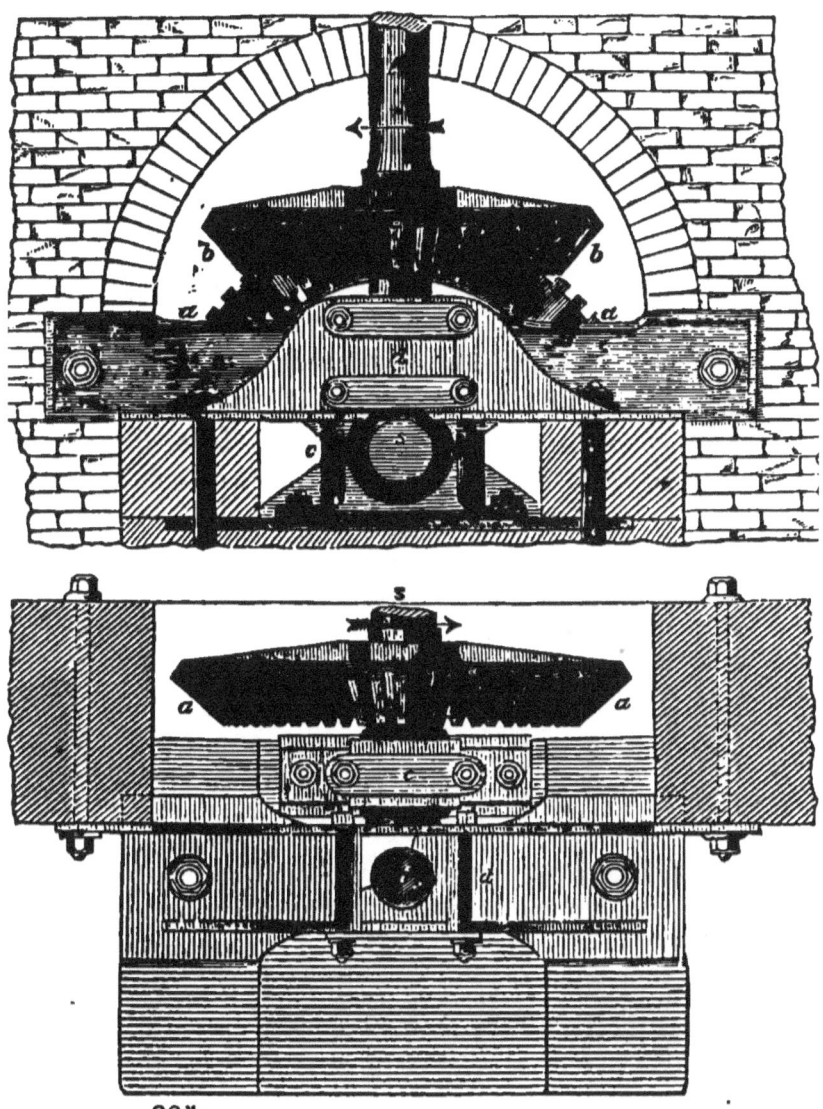

bevel wheel on the latter; *c* a plummer-block for the first motion-shaft, and *d d* the box containing the brass footstep for the vertical shaft; this box rests on a large base plate, bolted to the foundation stones and to the wall of the engine-house. In order to insure a constant supply of oil to the bearing, it is usual to cut away nearly the whole depth of the footstep, or that portion of the brass in the corner opposite to the thrust of the bevel wheels, as shown in the plan, fig. 154; this cavity is then kept full of oil, and lubricates the shaft throughout at every revolution.* Again, in cotton, woolen, and flax mills, when the first motion and vertical shafts have been duly proportioned to the work they have to perform, it becomes necessary to consider the diameter, speeds, etc., of the light shafting for driving the machinery in the different rooms. The formula given for strength, etc., in a former part of this work, will apply to this description of gearing and mill-work where the length does not exceed 120 feet. In long ranges of shafts, of from 150 to 200 feet in length, where the power applied to the machinery at the end of the room is considerable, it is essentially necessary to increase their strengths in order to prevent torsion or twist. This is a consideration of much importance, and requires careful

---

* The reader may compare what is here said of footsteps with that in Mills and Millwork, Part I., pp. 168, 172, on the steps for turbine shafts.

attention, as long ranges of light shafts are very elastic—they, in some cases, effect nearly a complete revolution at the point of imparted motion before the extreme ends begin to move. The result of the power so irregularly transmitted by the spring of the shafts, resolves itself into a series of accelerated and retarded motions through the whole line of shafts, and imparts to the machinery in one-half of the room a very variable motion. Want of stiffness is a great evil in long lines of shafting, and, as we have already observed, instances are not wanting in which whole lines have been removed entirely from this cause.

The transmission of power to machinery placed at different angles from the line of shafts, which is sometimes the case in old mills, has generally been effected by the universal joint A, fig. 155,

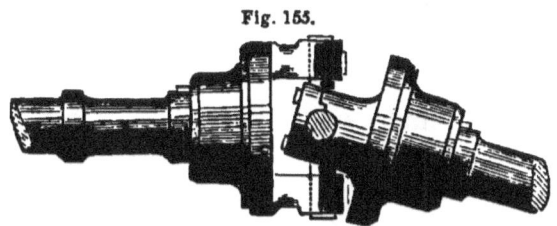

Fig. 155.

which works moderately well at an obtuse angle, but I have always found in my own practice that bevel wheels, as at B, fig. 156, are preferable and more satisfactory. They give much less trouble, and work with greater ease, than the universal joint. Other examples might be given for the guidance of the practical millwright; but, having to discuss these points at greater length when we come to treat of the different kinds of mills and different methods of gearing, we must direct the

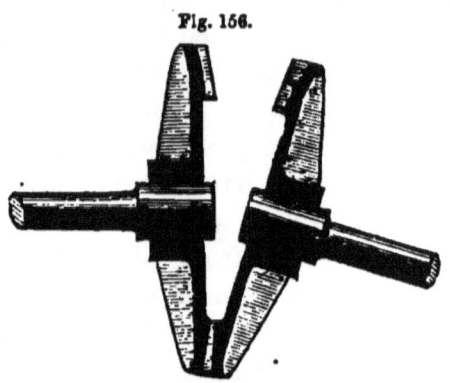

Fig. 156.

reader to those portions of the work which concern his own immediate practice.

The following table exhibits the diameter of shafts, length of journals, diameter and proportions of couplings, etc., derived from actual practice, which may be useful to the less experienced millwright and engineer:

TABLE 10.—LENGTH, DIAMETER, ETC., OF COUPLINGS, COUPLING-BOXES, ETC.

Diameter of Shaft.	Length of Neck.	Diameter of Coupling.	Length of Lap.	Length of Box.	Diameter of Box.	Thickness of Metal.
& 1⅝	3	2½	2	4½	4¼	⅞
1¾	3¼	3	2¼	5	5	1
2	4	3¼	2¼	5½	5½	1⅛
2¼	4½	3½	2¾	6	6	1¼
2½	5	4	3	6½	6¾	1⅜
2¾	5	4½	3¼	7	7½	1½
3	5½	4¾	3½	7½	7¾	1½
3¼	6¼	5	3¾	8	8¼	1⅝
3½	6½	5½	3¾	8¼	8¾	1⅝
4	7	6	4	8½	9½	1¾
4½	7½	6½	4½	9	10½	2
5	8	7¼	5	10	11¼	2
5½	8½	8	5½	11	12¼	2⅛
6	9	9	6	12	13½	2¼
6½	9½	9¾	6½	13	14¾	2½
7	10¼	10½	7	14	16	2¾
7½	11¼	11¼	7½	15	17	2⅞
8	12	12	8	16½	18	3
8½	12½	12½	8½	17	19	3¼
9	13½	13	9	18	20	3½
9½	14	13½	9½	18	21	3¾
10	14½	14	10	18½	22¾	3¾
11	15	16	11	20	24	4
12	16	17½	12	21	26	4¼
13	17	18½	13	22	27½	4½

INDEX.

	PAGE
Annular wheels	64
Archimedean screw creeper	91
Beam, the great, and the crank	25
Bevel wheels and bevel gear	66
—— skew bevel wheels	160
—— bevel wheels preferable to universal joint	259
Cams	76
—— to find the curve forming the groove of a camb, so that the velocity ratio of the rod and axis of the camb may be constant	77
—— to produce a changing reciprocating motion by a combination of the camb and screw	94
Concentric wheels	64
Connectors, wrapping	40
Crank and great beam, the	25
—— relations of crank and piston	27
Crown wheels	65
Cutting machine of Messrs. Peter Fairbairn & Co., of Leeds	111
Detent, ratchet wheel and	39
Eccentric wheel, the	75
Epicycloidal teeth of wheels	125
Face wheel and lantern	65
—— wheels and trundles	108
Fairbairn, Messrs. Peter & Co., of Leeds, their cutting machine	111
Friction of shafting	209
—— clutch	226
—— cones	227
—— coupling	228
—— discs	229
—— means adopted to lessen the, at the foot of the main vertical shaft	255
Guide pulleys	47
Gutta percha, value of, for straps	101
Hangers	238

	PAGE
Hero of Alexandria, his mention of toothed wheels	104
Idle wheels	63
Intermittent motion produced by linkwork, connected with a ratchet-wheel	39
Involute teeth of wheels	135
Journals, length of	207
Lantern, face-wheel and	65
Link-work	22
—— the crank and great beam	25
—— to construct Watt's parallel motion	31
—— to multiply oscillations by means of link-work	34
—— to produce a velocity which shall be rapidly retarded by means of link-work	36
—— to produce a reciprocating intermittent motion by means of link-work	37
—— ratchet-wheel and detent	39
—— intermittent motion produced by link-work connected with a ratchet-wheel	39
Lubrication of shafting	213
Mechanism, principles of	13
—— general views relative to machines	13
—— definitions and preliminary expositions	13
—— parts of a machine	18
—— elementary forms of mechanism	21
—— link-work	22
—— the crank and great beam	25
—— to construct Watt's parallel motion	31
—— to multiply oscillations by means of link-work	34
—— to produce a velocity which shall be rapidly retarded by means of link-work	36
—— to produce a reciprocating intermittent motion by means of link-work	37

(263)

INDEX.

Mechanism—*continued*.
— ratchet wheel and detent....... 89
— intermittent motion produced by link-work connected with a ratchet-wheel.......................... 89
— wrapping connectors............. 40
— endless cord or belt............. 40
— speed pulleys...................... 44
— guide pulleys....................... 47
— to prevent wrapping connectors from slipping.................. 48
— systems of pulleys............... 51
— to produce a varying velocity ratio by means of wrapping connectors 54
— on wheel-work producing motion by rolling contact when the axes of motion are parallel.. 56
— idle wheels........................... 63
— annular wheels..................... 64
— concentric wheels................. 64
— wheel-work when the axes are not parallel to each other... 64
— face-wheel and lantern......... 65
— crown-wheels....................... 65
— to construct bevel wheels and bevel gear, when the axes are in the same plane................... 66
— to construct bevel gear when the axes are not in the same plane 68
— variable motions produced by wheel-work having rolling contact.. 69
— Roemer's wheels................... 71
— intermittent and reciprocating motions, produced by wheel-work having rolling contact.... 71
— the rack and pinion.............. 72
— sliding pieces, producing motion by sliding contact............ 74
— the wedge, or movable inclined plane........................... 74
— the eccentric wheel.............. 75
— cambs, wipers, and tappets... 76
— the swash plate..................... 80
— screws, different forms of..... 82
— for cutting screws................ 85
— to produce a changing reciprocating rectilinear motion by a combination of the camb and screw...................................... 94
— to produce a boring motion by a combination of the screw and toothed wheels............... 95
Machinery of transmission, on...... 99

Oscillations, to multiply by means of link-work 34
Odontograph, Prof. Willis's........... 142

Parallel motion, Watt's, to construct 31
Piston, relations of the crank and.. 27
Press, the common...................... 86
Pulleys, speed............................. 44
— guide 47
Pulleys, systems of..................... 51
— and wheels............................ 99
Pitch of wheels........................... 116
Pinion from Ramelli.................... 104
Plummer-blocks.......................... 238

Rack and pinion, the................... 72
Ratchet-wheel and detent, the..... 89
— intermittent motion produced by link-work connected with a ratchet-wheel 89
Reciprocating intermittent motion, to produce a, by means of link-work 37
Rolling contact, motion produced by 56
— variable motions produced by wheel-work having rolling contact.. 71
Ramelli, pinion from.................... 104
Rannie, Mr. John, his introduction of cast-iron into all the details of mill-work 111

Screws 82
— construction of a helix or screw 82
— pitch of a screw 83
— transmission of motion by the screw 83
— solid screw and nut.............. 85
— the common press............... 86
— compound screw................. 88
— endless screw...................... 89
— differential screw 90
— Archimedean screw creeper... 91
— mechanism for cutting screws 92
— to produce a changing reciprocating rectilinear motion by a combination of the camb and screw...................................... 94
— to produce a boring motion by a combination of the screw and toothed wheels............... 95
Sliding-pieces, producing motion by sliding contact................... 74
— the wedge, or movable inclined plane........................... 74
— the eccentric wheel 75
— cambs, wipers, and tappets... 76
— the swash plate 80
— screws.................................. 82
— the common press............... 86

INDEX.

Sliding pieces—*continued.*
— to produce a changing reciprocating motion by a combination of the camb and screw 94
Speed pulleys 44
Swash plate, the 80
Saltaire mills, the 252
Shafts, on the strength and proportion of 175
— 1. The Material of which shafting is constructed 177
— 2. Transverse Strain 179
— rules for the strength of shafts 183
— resistance to flexure; weights producing a deflection of 1-1200 of the length in cast-iron cylindrical shafts 187
— resistance to flexure; weights producing a deflection of 1-1200 of the length in wrought-iron cylindrical shafts 188
— deflection arising from the weight of the shaft in both cast-iron and wrought-iron cylindrical shafts 189
— 3. Torsion 190
— values of modulus of torsion according to Mr. Bevan 194
— resume of experiments on cylinders of circular section 196
— resume of experiments on the torsion of hollow cylinders of copper 197
— resume of experiments on the torison of elliptical bars 197
— safe working torsion for cast-iron shafts 200
— safe working torsion for wrought-iron shafts 201
— diameter of wrought-iron shafting necessary to transmit with safety various amounts of force 205
— 4. Velocity of Shafts 204
— 5. Length of Journals 207
— 6. Friction 209
— table of coefficients of friction under pressures increased continually up to limits of abrasion 212
— 7. Lubrication 213
— 8. On Couplings for shafts and engaging and disengaging gear.. 216
— disengaging and re-engaging gear 221
— Hangers, Plummer-blocks, etc., for carrying shafting 238
— diameters, pitch, velocity, etc., of spur fly-wheels of the new construction 250

Shafts—*continued.*
— Material, etc., of the main shafts 251
— vertical shafts 252
— the Saltaire mills 252
— table of length, diameter, etc., of couplings, coupling-boxes, etc. 261
Skew bevel-wheels 160
Smeaton's introduction of cast-iron gearing in place of wood 107
Spur gearing 114
Straps compared with geared wheel-work 100
— materials of which straps are made 101
— strength of straps 102
— table of the least width of straps for transmitting various amounts of work over different pulleys 103

Tappets, or wipers 76
Tables relating to straps. *See* Wrapping connectors:
— wheels. *See* Wheel-work.
— shafts. *See* Shafts.
Teeth of wheels. *See* Wheel-work.
Toothed wheels, history of 103

Universal joint, bevel wheels preferable to 259

Velocity, to produce a, which shall be rapidly retarded, by means of link-work 36

Watt's parallel motion, to construct 31
Wedge, the, or movable inclined plane 74
Wheel-work, producing motion of, by rolling contact when the axes of motion are parallel 56
— idle-wheels 63
— annular wheels 64
— concentric wheels 64
— when the axes are not parallel to each other 64
— face-wheel and lantern 65
— crown wheels 65
— to construct bevel wheels, or bevel gear, when the axes are in the same plane 66
— to construct bevel gear when the axes are not in the same plane 68
— variable motions produced by wheel-work having rolling contact 69
— Roemer's wheels 71

INDEX

Wheel-work—*continued*.
—— intermittent and reciprocating motions, produced by wheel-work having rolling contact...... 71
—— to produce a boring motion by a combination of the screw and toothed wheels.................. 95
Wipers, or tappets................... 76
Wrapping connectors 40
—— endless cord or belt............. 40
—— speed pulleys..................... 44
—— guide pulleys..................... 47
—— to prevent wrapping connectors from slipping................ 48
—— systems of pulleys.............. 51
—— to produce a varying velocity ratio by means of wrapping connectors.......................... 54
Wheel-work, power of straps compared with that of geared wheel-work............................ 100
—— history of toothed wheels..... 103
—— Hero of Alexandria............. 104
—— Ramelli........................... 104
—— introduction of cast-iron gearing................................. 107
—— face-wheels and trundles...... 108
—— bevel wheels...................... 66
—— causes of the defects of wheel-work................................. 109
—— cutting machine of Messrs. Peter Fairbairn & Co., of Leeds.. 111
—— definitions of spur gearing.... 114
—— the pitch of wheels............. 116
—— table of the relation of diameter, pitch, and number of teeth, for wheels of from ½ inch to 5 inches pitch, and from 12 to 200 teeth................................... 123
—— the principles which determine the proper form of the teeth of wheels..................... 124
—— construction of epicycloidal teeth................................... 129

Wheel-work—*continued*.
—— epicycloidal teeth................ 125
—— the rack........................... 135
—— involute teeth.................... 135
—— Prof. Willis's method of striking the teeth of wheels............ 140
—— Prof. Willis, his odontograph. 142
—— general form and proportion of toothed wheels.................. 145
—— table giving the proportions of the teeth of wheels in inches and thirty seconds of an inch..... 156
—— table of proportions of teeth of wheels for average practice... 154
—— skew bevel wheels................ 160
—— worm and wheel................... 163
—— strength of the teeth of wheels................................. 165
—— table of thickness, breadth and pitch of teeth of wheels..... 168
—— table of the relation of horses' power transmitted, and velocity at the pitch circle, to pressure on teeth.............................. 172
—— table showing the pitch and thickness of teeth to transmit a given number of horses' power at different velocities............. 173
—— table showing the breadth of teeth required to transmit different amounts of force at a uniform pressure of 400 lbs per inch. 174
Wheels and pulleys..................... 99
Willis, Prof., his method of striking the teeth of wheels................ 140
—— his odontograph................... 142
Worm and wheel, the................... 163
Wrapping connectors.................... 99
—— power of straps compared with that of geared wheel-work. 100
—— table of the approximate width of leather straps in inches necessary to transmit any number of horses' power................ 103

CATALOGUE
OF
PRACTICAL AND SCIENTIFIC BOOKS,
PUBLISHED BY
HENRY CAREY BAIRD,
INDUSTRIAL PUBLISHER,
No. 406 WALNUT STREET,
PHILADELPHIA.

☞ Any of the Books comprised in this Catalogue will be sent by mail, free of postage, at the publication price.

☞ This Catalogue will be sent, free of postage, to any one who will furnish the publisher with his address.

ARMENGAUD, AMOUROUX, AND JOHNSON.—THE PRACTICAL DRAUGHTSMAN'S BOOK OF INDUSTRIAL DESIGN, AND MACHINIST'S AND ENGINEER'S DRAWING COMPANION: Forming a complete course of Mechanical Engineering and Architectural Drawing. From the French of M. Armengaud the elder, Prof. of Design in the Conservatoire of Arts and Industry, Paris, and MM. Armengaud the younger and Amouroux, Civil Engineers. Rewritten and arranged, with additional matter and plates, selections from and examples of the most useful and generally employed mechanism of the day. By WILLIAM JOHNSON, Assoc. Inst. C. E., Editor of "The Practical Mechanic's Journal." Illustrated by 50 folio steel plates and 50 wood-cuts. A new edition, 4to. . $10 00

ARROWSMITH.—PAPER-HANGER'S COMPANION: A Treatise in which the Practical Operations of the Trade are Systematically laid down: with Copious Directions Preparatory to Papering; Preventives against the Effect of Damp on Walls; the Various Cements and Pastes adapted to the Several Purposes of the Trade; Observations and Directions for the Panelling and Ornamenting of Rooms, &c. By JAMES ARROWSMITH, Author of "Analysis of Drapery," &c. 12mo., cloth $1 25

BAIRD.—THE AMERICAN COTTON SPINNER, AND MANAGER'S AND CARDER'S GUIDE:

A Practical Treatise on Cotton Spinning; giving the Dimensions and Speed of Machinery, Draught and Twist Calculations, etc.; with notices of recent Improvements: together with Rules and Examples for making changes in the sizes and numbers of Roving and Yarn. Compiled from the papers of the late ROBERT H. BAIRD. 12mo. . . . $1 50

BAKER.—LONG-SPAN RAILWAY BRIDGES:

Comprising Investigations of the Comparative Theoretical and Practical Advantages of the various Adopted or Proposed Type Systems of Construction; with numerous Formulæ and Tables. By B. Baker. 12mo. $2 00

BAKEWELL.—A MANUAL OF ELECTRICITY—PRACTICAL AND THEORETICAL:

By F. C. BAKEWELL, Inventor of the Copying Telegraph. Second Edition. Revised and enlarged. Illustrated by numerous engravings. 12mo. Cloth $2 00

BEANS.—A TREATISE ON RAILROAD CURVES AND THE LOCATION OF RAILROADS:

By E. W. BEANS, C. E. 12mo. (In press.)

BLENKARN.—PRACTICAL SPECIFICATIONS OF WORKS EXECUTED IN ARCHITECTURE, CIVIL AND MECHANICAL ENGINEERING, AND IN ROAD MAKING AND SEWERING:

To which are added a series of practically useful Agreements and Reports. By JOHN BLENKARN. Illustrated by fifteen large folding plates. 8vo. $9 00

BLINN.—A PRACTICAL WORKSHOP COMPANION FOR TIN, SHEET-IRON, AND COPPER-PLATE WORKERS:

Containing Rules for Describing various kinds of Patterns used by Tin, Sheet-iron, and Copper-plate Workers; Practical Geometry; Mensuration of Surfaces and Solids; Tables of the Weight of Metals, Lead Pipe, etc.; Tables of Areas and Circumferences of Circles; Japans, Varnishes, Lackers, Cements, Compositions, etc. etc. By LEROY J. BLINN, Master Mechanic. With over One Hundred Illustrations. 12mo. $2 50

BOOTH.—MARBLE WORKER'S MANUAL:
Containing Practical Information respecting Marbles in general, their Cutting, Working, and Polishing; Veneering of Marble; Mosaics; Composition and Use of Artificial Marble, Stuccos, Cements, Receipts, Secrets, etc. etc. Translated from the French by M. L. BOOTH. With an Appendix concerning American Marbles. 12mo., cloth . . $1 50

BOOTH AND MORFIT.—THE ENCYCLOPEDIA OF CHEMISTRY, PRACTICAL AND THEORETICAL:
Embracing its application to the Arts, Metallurgy, Mineralogy, Geology, Medicine, and Pharmacy. By JAMES C. BOOTH, Melter and Refiner in the United States Mint, Professor of Applied Chemistry in the Franklin Institute, etc., assisted by CAMPBELL MORFIT, author of "Chemical Manipulations," etc. Seventh edition. Complete in one volume, royal 8vo., 978 pages, with numerous wood-cuts and other illustrations. $5 00

BOWDITCH.—ANALYSIS, TECHNICAL VALUATION, PURIFICATION, AND USE OF COAL GAS:
By Rev. W. R. BOWDITCH. Illustrated with wood engravings. 8vo. $6 50

BOX.—PRACTICAL HYDRAULICS:
A Series of Rules and Tables for the use of Engineers, etc. By THOMAS BOX. 12mo. $2 00

BUCKMASTER.—THE ELEMENTS OF MECHANICAL PHYSICS:
By J. C. BUCKMASTER, late Student in the Government School of Mines; Certified Teacher of Science by the Department of Science and Art; Examiner in Chemistry and Physics in the Royal College of Preceptors; and late Lecturer in Chemistry and Physics of the Royal Polytechnic Institute. Illustrated with numerous engravings. In one vol. 12mo. . $2 00

BULLOCK.—THE AMERICAN COTTAGE BUILDER:
A Series of Designs, Plans, and Specifications, from $200 to to $20,000 for Homes for the People; together with Warming, Ventilation, Drainage, Painting, and Landscape Gardening. By JOHN BULLOCK, Architect, Civil Engineer, Mechanician, and Editor of "The Rudiments of Architecture and Building," etc. Illustrated by 75 engravings. In one vol. 8vo. $3 50

BULLOCK.— THE RUDIMENTS OF ARCHITECTURE AND BUILDING:
For the use of Architects, Builders, Draughtsmen, Machinists, Engineers, and Mechanics. Edited by John Bullock, author of "The American Cottage Builder." Illustrated by 250 engravings. In one volume 8vo. . . . $3 50

BURGH.—PRACTICAL ILLUSTRATIONS OF LAND AND MARINE ENGINES:
Showing in detail the Modern Improvements of High and Low Pressure, Surface Condensation, and Super-heating, together with Land and Marine Boilers. By N. P. Burgh, Engineer. Illustrated by twenty plates, double elephant folio, with text.
$21 00

BURGH.—PRACTICAL RULES FOR THE PROPORTIONS OF MODERN ENGINES AND BOILERS FOR LAND AND MARINE PURPOSES.
By N. P. Burgh, Engineer. 12mo. . . . $2 00

BURGH.—THE SLIDE-VALVE PRACTICALLY CONSIDERED:
By N. P. Burgh, author of "A Treatise on Sugar Machinery," "Practical Illustrations of Land and Marine Engines," "A Pocket-Book of Practical Rules for Designing Land and Marine Engines, Boilers," etc. etc. etc. Completely illustrated. 12mo. $2 00

BYRN.—THE COMPLETE PRACTICAL BREWER:
Or, Plain, Accurate, and Thorough Instructions in the Art of Brewing Beer, Ale, Porter, including the Process of making Bavarian Beer, all the Small Beers, such as Root-beer, Gingerpop, Sarsaparilla-beer, Mead, Spruce beer, etc. etc. Adapted to the use of Public Brewers and Private Families. By M. La Fayette Byrn, M. D. With illustrations. 12mo. $1 25

BYRN.—THE COMPLETE PRACTICAL DISTILLER:
Comprising the most perfect and exact Theoretical and Practical Description of the Art of Distillation and Rectification; including all of the most recent improvements in distilling apparatus; instructions for preparing spirits from the numerous vegetables, fruits, etc.; directions for the distillation and preparation of all kinds of brandies and other spirits, spirituous and other compounds, etc. etc.; all of which is so simplified that it is adapted not only to the use of extensive distillers, but for every farmer, or others who may wish to engage in the art of distilling. By M. La Fayette Byrn, M. D. With numerous engravings. In one volume, 12mo. $1 50

BYRNE.—POCKET BOOK FOR RAILROAD AND CIVIL ENGINEERS:
Containing New, Exact, and Concise Methods for Laying out Railroad Curves, Switches, Frog Angles and Crossings; the Staking out of work; Levelling; the Calculation of Cuttings; Embankments; Earth-work, etc. By OLIVER BYRNE. Illustrated, 18mo. $1 25

BYRNE.—THE HANDBOOK FOR THE ARTISAN, MECHANIC, AND ENGINEER:
By OLIVER BYRNE. Illustrated by 11 large plates and 185 Wood Engravings. 8vo. $5 00

BYRNE.—THE ESSENTIAL ELEMENTS OF PRACTICAL MECHANICS:
For Engineering Students, based on the Principle of Work. By OLIVER BYRNE. Illustrated by Numerous Wood Engravings, 12mo. $3 63

BYRNE.—THE PRACTICAL METAL-WORKER'S ASSISTANT:
Comprising Metallurgic Chemistry; the Arts of Working all Metals and Alloys; Forging of Iron and Steel; Hardening and Tempering; Melting and Mixing; Casting and Founding; Works in Sheet Metal; the Processes Dependent on the Ductility of the Metals; Soldering; and the most Improved Processes and Tools employed by Metal-Workers. With the Application of the Art of Electro-Metallurgy to Manufacturing Processes; collected from Original Sources, and from the Works of Holtzapffel, Bergeron, Leupold, Plumier, Napier, and others. By OLIVER BYRNE. A New, Revised, and improved Edition, with Additions by John Scoffern, M. B , William Clay, Wm. Fairbairn, F. R. S., and James Napier. With Five Hundred and Ninety-two Engravings; Illustrating every Branch of the Subject. In one volume, 8vo. 652 pages . $7 00

BYRNE.—THE PRACTICAL CALCULATOR:
For the Engineer, Mechanic, Manufacturer of Engine Work, Naval Architect, Miner, and Millwright. By OLIVER BYRNE. 1 volume, 8vo., nearly 600 pages $4 50

CABINET MAKER'S ALBUM OF FURNITURE:
Comprising a Collection of Designs for the Newest and Most Elegant Styles of Furniture. Illustrated by Forty eight Large and Beautifully Engraved Plates. In one volume, oblong
$5 00

CALVERT.—LECTURES ON COAL-TAR COLORS, AND ON RECENT IMPROVEMENTS AND PROGRESS IN DYEING AND CALICO PRINTING:

Embodying Copious Notes taken at the last London International Exhibition, and *Illustrated with Numerous Patterns of Aniline and other Colors.* By F. GRACE CALVERT, F. R. S., F. C. S., Professor of Chemistry at the Royal Institution, Manchester, Corresponding Member of the Royal Academies of Turin and Rouen; of the Pharmaceutical Society of Paris; Société Industrielle de Mulhouse, etc. In one volume, 8vo., cloth $1 50

CAMPIN.—A PRACTICAL TREATISE ON MECHANICAL ENGINEERING:

Comprising Metallurgy, Moulding, Casting, Forging, Tools, Workshop Machinery, Mechanical Manipulation, Manufacture of Steam-engines, etc. etc. With an Appendix on the Analysis of Iron and Iron Ores. By FRANCIS CAMPIN, C. E. To which are added, Observations on the Construction of Steam Boilers, and Remarks upon Furnaces used for Smoke Prevention; with a Chapter on Explosions. By R. Armstrong, C. E., and John Bourne. Rules for Calculating the Change Wheels for Screws on a Turning Lathe, and for a Wheel-cutting Machine. By J. LA NICCA. Management of Steel, including Forging, Hardening, Tempering, Annealing, Shrinking, and Expansion. And the Case-hardening of Iron. By G. EDE. 8vo. Illustrated with 29 plates and 100 wood engravings.
$6 00

CAMPIN.—THE PRACTICE OF HAND-TURNING IN WOOD, IVORY, SHELL, ETC.:

With Instructions for Turning such works in Metal as may be required in the Practice of Turning Wood, Ivory, etc. Also, an Appendix on Ornamental Turning. By FRANCIS CAMPIN; with Numerous Illustrations, 12mo., cloth . . $3 00

CAPRON DE DOLE.—DUSSAUCE.—BLUES AND CARMINES OF INDIGO.

A Practical Treatise on the Fabrication of every Commercial Product derived from Indigo. By FELICIEN CAPRON DE DOLE. Translated, with important additions, by Professor H. DUSSAUCE. 12mo. $2 50

CAREY.—THE WORKS OF HENRY C. CAREY:
CONTRACTION OR EXPANSION? REPUDIATION OR RESUMPTION? Letters to Hon. Hugh McCulloch. 8vo. 38
FINANCIAL CRISES, their Causes and Effects. 8vo. paper 25
HARMONY OF INTERESTS; Agricultural, Manufacturing, and Commercial. 8vo., paper $1 00
Do. do. cloth . . . $1 50
LETTERS TO THE PRESIDENT OF THE UNITED STATES. Paper 75
MANUAL OF SOCIAL SCIENCE. Condensed from Carey's "Principles of Social Science." By KATE MCKEAN. 1 vol. 12mo. $2 25
MISCELLANEOUS WORKS: comprising "Harmony of Interests," "Money," "Letters to the President," "French and American Tariffs," "Financial Crises," "The Way to Outdo England without Fighting Her," "Resources of the Union," "The Public Debt," "Contraction or Expansion," "Review of the Decade 1857—'67," "Reconstruction," etc. etc. 1 vol. 8vo., cloth $4 50
MONEY: A LECTURE before the N. Y. Geographical and Statistical Society. 8vo., paper 25
PAST, PRESENT, AND FUTURE. 8vo. . . . $2 50
PRINCIPLES OF SOCIAL SCIENCE. 3 volumes 8vo., cloth $10 00
REVIEW OF THE DECADE 1857—'67. 8vo., paper 38
RECONSTRUCTION: INDUSTRIAL, FINANCIAL, AND POLITICAL. Letters to the Hon. Henry Wilson, U. S. S. 8vo. paper 38
THE PUBLIC DEBT, LOCAL AND NATIONAL. How to provide for its discharge while lessening the burden of Taxation. Letter to David A. Wells, Esq., U. S. Revenue Commission. 8vo., paper 25
THE RESOURCES OF THE UNION. A Lecture read, Dec. 1865, before the American Geographical and Statistical Society, N. Y., and before the American Association for the Advancement of Social Science, Boston . . . 25
THE SLAVE TRADE, DOMESTIC AND FOREIGN; Why it Exists, and How it may be Extinguished. 12mo., cloth $1 50

THE WAY TO OUTDO ENGLAND WITHOUT FIGHTING HER. Letters to the Hon. Schuyler Colfax, Speaker of the House of Representatives United States, on "The Paper Question," "The Farmer's Question," "The Iron Question," "The Railroad Question," and "The Currency Question." 8vo., paper 75

CHEVALIER.—THE PHOTOGRAPHIC STUDENT.
A Complete Treatise on the Theory and Practice of Photography. Translated from the French of A. CHEVALIER. Illustrated by numerous engravings. (In press.)

CLOUGH.—THE CONTRACTOR'S MANUAL AND BUILDER'S PRICE-BOOK:
Designed to elucidate the method of ascertaining, correctly, the value and Quantity of every description of Work and Materials used in the Art of Building, from their Prime Cost in any part of the United States, collected from extensive experience and observation in Building and Designing; to which are added a large variety of Tables, Memoranda, etc., indispensable to all engaged or concerned in erecting buildings of any kind. By A. B. CLOUGH, Architect, 24mo., cloth 75

COLBURN.—THE GAS-WORKS OF LONDON:
Comprising a sketch of the Gas-works of the city, Process of Manufacture, Quantity Produced, Cost, Profit, etc. By ZERAH COLBURN. 8vo., cloth 75

COLBURN.—THE LOCOMOTIVE ENGINE:
Including a Description of its Structure, Rules for Estimating its Capabilities, and Practical Observations on its Construction and Management. By ZERAH COLBURN. Illustrated. A new edition. 12mo. $1 25

COLBURN AND MAW.—THE WATER-WORKS OF LONDON:
Together with a Series of Articles on various other Waterworks. By ZERAH COLBURN and W. MAW. Reprinted from "Engineering." In one volume, 8vo. . . $1 00

DAGUERREOTYPIST AND PHOTOGRAPHER'S COMPANION:
12mo., cloth $1 25

DAVIS.—A TREATISE ON HARNESS, SADDLES, AND BRIDLES:
Their History and Manufacture from the Earliest Times down to the Present Period. By A. DAVIS, Practical Saddler and Harness Maker. (In press.)

DESSOYE.—STEEL, ITS MANUFACTURE, PROPERTIES, AND USE.
By J. B. J. Dessoye, Manufacturer of Steel; with an Introduction and Notes by Ed. Graten, Engineer of Mines. Translated from the French. In one volume, 12mo. (In press.)

DIRCKS.—PERPETUAL MOTION:
Or Search for Self-Motive Power during the 17th, 18th, and 19th centuries. Illustrated from various authentic sources in Papers, Essays, Letters, Paragraphs, and numerous Patent Specifications, with an Introductory Essay by Henry Dircks, C. E. Illustrated by numerous engravings of machines. 12mo., cloth $3 50

DIXON.—THE PRACTICAL MILLWRIGHT'S AND ENGINEER'S GUIDE:
Or Tables for Finding the Diameter and Power of Cogwheels; Diameter, Weight, and Power of Shafts; Diameter and Strength of Bolts, etc. etc. By Thomas Dixon. 12mo., cloth. $1 50

DUNCAN.—PRACTICAL SURVEYOR'S GUIDE:
Containing the necessary information to make any person, of common capacity, a finished land surveyor without the aid of a teacher. By Andrew Duncan. Illustrated. 12mo., cloth. $1 25

DUSSAUCE.—A NEW AND COMPLETE TREATISE ON THE ARTS OF TANNING, CURRYING, AND LEATHER DRESSING:
Comprising all the Discoveries and Improvements made in France, Great Britain, and the United States. Edited from Notes and Documents of Messrs. Sallerou, Grouvelle, Duval, Dessables, Labarraque, Payen, René, De Fontenelle, Malapeyre, etc. etc. By Prof. H. Dussauce, Chemist. Illustrated by 212 wood engravings. 8vo. $10 00

DUSSAUCE.—A GENERAL TREATISE ON THE MANUFACTURE OF EVERY DESCRIPTION OF SOAP:
Comprising the Chemistry of the Art, with Remarks on Alkalies, Saponifiable Fatty Bodies, the apparatus necessary in a Soap Factory, Practical Instructions on the manufacture of the various kinds of Soap, the assay of Soaps, etc. etc. Edited from notes of Larmé, Fontenelle, Malapeyre, Dufour, and others, with large and important additions by Professor H. Dussauce, Chemist. Illustrated. In one volume, 8vo. (In press.)

DUSSAUCE.—A PRACTICAL GUIDE FOR THE PERFUMER:
Being a New Treatise on Perfumery the most favorable to the Beauty without being injurious to the Health, comprising a Description of the substances used in Perfumery, the Formulœ of more than one thousand Preparations, such as Cosmetics, Perfumed Oils, Tooth Powders, Waters, Extracts, Tinctures, Infusions, Vinaigres, Essential Oils, Pastels, Creams, Soaps, and many new Hygienic Products not hitherto described. Edited from Notes and Documents of Messrs. Debay, Lunel, etc. With additions by Professor H. DUSSAUCE, Chemist. (In press, *shortly to be issued.*)

DUSSAUCE.—PRACTICAL TREATISE ON THE FABRICATION OF MATCHES, GUN COTTON, AND FULMINATING POWDERS.
By Professor H. DUSSAUCE. 12mo. . . . $3 00

DUSSAUCE.—TREATISE ON THE COLORING MATTERS DERIVED FROM COAL TAR:
Their Practical Application in Dyeing Cotton, Wool, and Silk; the Principles of the Art of Dyeing and of the Distillation of Coal Tar, with a Description of the most Important New Dyes now in use. By Prof. H. DUSSAUCE. 12mo. . $3 00

DYER AND COLOR-MAKER'S COMPANION:
Containing upwards of two hundred Receipts for making Colors, on the most approved principles, for all the various styles and fabrics now in existence; with the Scouring Process, and plain Directions for Preparing, Washing-off, and Finishing the Goods. In one vol. 12mo. $1 25

EASTON.—A PRACTICAL TREATISE ON STREET OR HORSE-POWER RAILWAYS:
Their Location, Construction, and Management; with General Plans and Rules for their Organization and Operation; together with Examinations as to their Comparative Advantages over the Omnibus System, and Inquiries as to their Value for Investment; including Copies of Municipal Ordinances relating thereto. By ALEXANDER EASTON, C. E. Illustrated by 23 plates, 8vo., cloth $2 00

ERNI.—COAL OIL AND PETROLEUM:
Their Origin, History, Geology, and Chemistry; with a view of their importance in their bearing on National Industry. By Dr. HENRI ERNI, Chief Chemist, Department of Agriculture. 12mo. $2 50

ERNI.—THE THEORETICAL AND PRACTICAL CHEMISTRY OF FERMENTATION:
Comprising the Chemistry of Wine, Beer, Distilling of Liquors; with the Practical Methods of their Chemical Examination, Preservation, and Improvement—such as Gallizing of Wines. With an Appendix, containing well-tested Practical Rules and Receipts for the manufacture, etc., of all kinds of Alcoholic Liquors. By Henry Erni, Chief Chemist, Department of Agriculture. (In press.)

FAIRBAIRN.—THE PRINCIPLES OF MECHANISM AND MACHINERY OF TRANSMISSION:
Comprising the Principles of Mechanism, Wheels, and Pulleys, Strength and Proportions of Shafts, Couplings of Shafts, and Engaging and Disengaging Gear. By William Fairbairn, Esq., C. E., LL. D., F. R. S., F. G. S., Corresponding Member of the National Institute of France, and of the Royal Academy of Turin; Chevalier of the Legion of Honor, etc. etc. Beautifully illustrated by over 150 wood-cuts. In one volume 12mo. $2 50

FAIRBAIRN.—PRIME-MOVERS:
Comprising the Accumulation of Water-power; the Construction of Water-wheels and Turbines; the Properties of Steam; the Varieties of Steam-engines and Boilers and Wind-mills. By William Fairbairn, C. E., LL. D., F. R. S., F. G. S. Author of "Principles of Mechanism and the Machinery of Transmission." With Numerous Illustrations. In one volume. (In press.)

FLAMM.—A PRACTICAL GUIDE TO THE CONSTRUCTION OF ECONOMICAL HEATING APPLICATIONS FOR SOLID AND GASEOUS FUELS:
With the Application of Concentrated Heat, and on Waste Heat, for the Use of Engineers, Architects, Stove and Furnace Makers, Manufacturers of Fire Brick, Zinc, Porcelain, Glass, Earthenware, Steel, Chemical Products, Sugar Refiners, Metallurgists, and all others employing Heat. By M. Pierre Flamm, Manufacturer. Illustrated. Translated from the French. One volume, 12mo. (In press.)

GILBART.—A PRACTICAL TREATISE ON BANKING:
By James William Gilbart. To which is added: The National Bank Act as now (1868) in force. 8vo. $4 50

GOTHIC ALBUM FOR CABINET MAKERS:
Comprising a Collection of Designs for Gothic Furniture. Illustrated by twenty-three large and beautifully engraved plates. Oblong $3 00

GRANT.—BEET-ROOT SUGAR AND CULTIVATION OF THE BEET:
By E. B. GRANT. 12mo. $1 25

GREGORY.—MATHEMATICS FOR PRACTICAL MEN:
Adapted to the Pursuits of Surveyors, Architects, Mechanics, and Civil Engineers. By OLINTHUS GREGORY. 8vo., plates, cloth $3 00

GRISWOLD.—RAILROAD ENGINEER'S POCKET COMPANION.
Comprising Rules for Calculating Deflection Distances and Angles, Tangential Distances and Angles, and all Necessary Tables for Engineers; also the art of Levelling from Preliminary Survey to the Construction of Railroads, intended Expressly for the Young Engineer, together with Numerous Valuable Rules and Examples. By W. GRISWOLD. 12mo., tucks. $1 25

GUETTIER.—METALLIC ALLOYS:
Being a Practical Guide to their Chemical and Physical Properties, their Preparation, Composition, and Uses. Translated from the French of A. GUETTIER, Engineer and Director of Founderies, author of "La Fouderie en France," etc. etc. By A. A. FESQUET, Chemist and Engineer. In one volume, 12mo. (In press, *shortly to be published.*)

HATS AND FELTING:
A Practical Treatise on their Manufacture. By a Practical Hatter. Illustrated by Drawings of Machinery, &c., 8vo.

HAY.—THE INTERIOR DECORATOR:
The Laws of Harmonious Coloring adapted to Interior Decorations: with a Practical Treatise on House-Painting. By D. R. HAY, House-Painter and Decorator. Illustrated by a Diagram of the Primary, Secondary, and Tertiary Colors. 12mo. $2 25

HUGHES.—AMERICAN MILLER AND MILLWRIGHT'S ASSISTANT:
By WM. CARTER HUGHES. A new edition. In one volume, 12mo. $1 50

HUNT.—THE PRACTICE OF PHOTOGRAPHY.
By ROBERT HUNT, Vice-President of the Photographic Society, London, with numerous illustrations. 12mo., cloth . 75

HURST.—A HAND-BOOK FOR ARCHITECTURAL SURVEYORS:
Comprising Formulæ useful in Designing Builder's work, Table of Weights, of the materials used in Building, Memoranda connected with Builders' work, Mensuration, the Practice of Builders' Measurement, Contracts of Labor, Valuation of Property, Summary of the Practice in Dilapidation, etc. etc. By J. F. HURST, C. E. 2d edition, pocket-book form, full bound $2 50

JERVIS.—RAILWAY PROPERTY:
A Treatise on the Construction and Management of Railways; designed to afford useful knowledge, in the popular style, to the holders of this class of property; as well as Railway Managers, Officers, and Agents. By JOHN B. JERVIS, late Chief Engineer of the Hudson River Railroad, Croton Aqueduct, &c. One vol. 12mo., cloth $2 00

JOHNSON.—A REPORT TO THE NAVY DEPARTMENT OF THE UNITED STATES ON AMERICAN COALS:
Applicable to Steam Navigation and to other purposes. By WALTER R. JOHNSON. With numerous illustrations. 607 pp. 8vo., half morocco $6 00

JOHNSON.—THE COAL TRADE OF BRITISH AMERICA:
With Researches on the Characters and Practical Values of American and Foreign Coals. By WALTER R. JOHNSON, Civil and Mining Engineer and Chemist. 8vo. . . . $2 00

JOHNSTON.—INSTRUCTIONS FOR THE ANALYSIS OF SOILS, LIMESTONES, AND MANURES.
By J. W. F. JOHNSTON. 12mo. 38

KEENE.—A HAND-BOOK OF PRACTICAL GAUGING,
For the Use of Beginners, to which is added A Chapter on Distillation, describing the process in operation at the Custom House for ascertaining the strength of wines. By JAMES B. KEENE, of H. M. Customs. 8vo. $1 25

KENTISH.—A TREATISE ON A BOX OF INSTRUMENTS,
And the Slide Rule; with the Theory of Trigonometry and Logarithms, including Practical Geometry, Surveying, Measuring of Timber, Cask and Malt Gauging, Heights, and Distances. By THOMAS KENTISH. In one volume. 12mo. . $1 25

KOBELL.—ERNI.—MINERALOGY SIMPLIFIED;
A short method of Determining and Classifying Minerals, by means of simple Chemical Experiments in the Wet Way. Translated from the last German Edition of F. Von Kobell, with an Introduction to Blowpipe Analysis and other additions. By Henri Erni, M. D., Chief Chemist, Department of Agriculture, author of "Coal Oil and Petroleum." In one volume, 12mo. $2 50

LAFFINEUR.—A PRACTICAL GUIDE TO HYDRAULICS FOR TOWN AND COUNTRY;
Or a Complete Treatise on the Building of Conduits for Water for Cities, Towns, Farms, Country Residences, Workshops, etc. Comprising the means necessary for obtaining at all times abundant supplies of Drinkable Water. Translated from the French of M. Jules Laffineur, C. E. Illustrated. (In press.)

LAFFINEUR.—A TREATISE ON THE CONSTRUCTION OF WATER-WHEELS:
Containing the various Systems in use with Practical Information on the Dimensions necessary for Shafts, Journals, Arms, etc., of Water-wheels, etc. etc. Translated from the French of M. Jules Laffineur, C. E. Illustrated by numerous plates. (In press.)

LANDRIN.—A TREATISE ON STEEL:
Comprising the Theory, Metallurgy, Practical Working, Properties, and Use. Translated from the French of H. C. Landrin, Jr., C. E. By A. A. Fesquet, Chemist and Engineer. Illustrated. 12mo. (In press.)

LARKIN.—THE PRACTICAL BRASS AND IRON FOUNDER'S GUIDE:
A Concise Treatise on Brass Founding, Moulding, the Metals and their Alloys, etc.; to which are added Recent Improvements in the Manufacture of Iron, Steel by the Bessemer Process, etc. etc. By James Larkin, late Conductor of the Brass Foundry Department in Reany, Neafie & Co.'s Penn Works, Philadelphia. Fifth edition, revised, with Extensive additions. In one volume, 12mo. $2 25

LEAVITT.—FACTS ABOUT PEAT AS AN ARTICLE OF FUEL:
With Remarks upon its Origin and Composition, the Localities in which it is found, the Methods of Preparation and Manufacture, and the various Uses to which it is applicable; together with many other matters of Practical and Scientific Interest. To which is added a chapter on the Utilization of Coal Dust with Peat for the Production of an Excellent Fuel at Moderate Cost, especially adapted for Steam Service. By H. T. LEAVITT. Third edition. 12mo. . . . $1 75

LEROUX.—A PRACTICAL TREATISE ON WOOLS AND WORSTEDS:
Translated from the French of CHARLES LEROUX, Mechanical Engineer, and Superintendent of a Spinning Mill. Illustrated by 12 large plates and 34 engravings. In one volume 8vo. (In press, *shortly to be published.*)

LESLIE (MISS).—COMPLETE COOKERY:
Directions for Cookery in its Various Branches. By MISS LESLIE. 58th thousand. Thoroughly revised, with the addition of New Receipts. In 1 vol. 12mo., cloth . . $1 25

LESLIE (MISS). LADIES' HOUSE BOOK:
a Manual of Domestic Economy. 20th revised edition. 12mo., cloth $1 25

LESLIE (MISS).—TWO HUNDRED RECEIPTS IN FRENCH COOKERY.
12mo. 50

LIEBER.—ASSAYER'S GUIDE:
Or, Practical Directions to Assayers, Miners, and Smelters, for the Tests and Assays, by Heat and by Wet Processes, for the Ores of all the principal Metals, of Gold and Silver Coins and Alloys, and of Coal, etc. By OSCAR M. LIEBER. 12mo., cloth
$1 25

LOVE.—THE ART OF DYEING, CLEANING, SCOURING, AND FINISHING:
On the most approved English and French methods; being Practical Instructions in Dyeing Silks, Woollens, and Cottons, Feathers, Chips, Straw, etc.; Scouring and Cleaning Bed and Window Curtains, Carpets, Rugs, etc.; French and English Cleaning, any Color or Fabric of Silk, Satin, or Damask. By THOMAS LOVE, a Working Dyer and Scourer. In 1 vol. 12mo.
$3 00

MAIN AND BROWN.—QUESTIONS ON SUBJECTS CONNECTED WITH THE MARINE STEAM-ENGINE:
And Examination Papers; with Hints for their Solution. By THOMAS J. MAIN, Professor of Mathematics, Royal Naval College, and THOMAS BROWN, Chief Engineer, R. N. 12mo., cloth
$1 50

MAIN AND BROWN.—THE INDICATOR AND DYNAMOMETER:
With their Practical Applications to the Steam-Engine. By THOMAS J. MAIN, M. A. F. R., Ass't Prof. Royal Naval College, Portsmouth, and THOMAS BROWN, Assoc. Inst. C. E., Chief Engineer, R. N., attached to the R. N. College. Illustrated. From the Fourth London Edition. 8vo. . . . $1 50

MAIN AND BROWN.—THE MARINE STEAM-ENGINE.
By THOMAS J. MAIN, F. R. Ass't S. Mathematical Professor at Royal Naval College, and THOMAS BROWN, Assoc. Inst. C. E. Chief Engineer, R. N. Attached to the Royal Naval College. Authors of "Questions connected with the Marine Steam-Engine," and the "Indicator and Dynamometer." With numerous Illustrations. In one volume, 8vo. . . . $5 00

MAKINS.—A MANUAL OF METALLURGY:
More particularly of the Precious Metals: including the Methods of Assaying them. Illustrated by upwards of 50 Engravings. By GEORGE HOGARTH MAKINS, M. R. C. S., F. C. S., one of the Assayers to the Bank of England, Assayer to the Anglo-Mexican Mints, and Lecturer upon Metallurgy at the Dental Hospital, London. In one volume, 12mo. . . $3 50

MARTIN —SCREW-CUTTING TABLES, FOR THE USE OF MECHANICAL ENGINEERS:
Showing the Proper Arrangement of Wheels for Cutting the Threads of Screws of any required Pitch; with a Table for Making the Universal Gas-Pipe Thread and Taps. By W. A. MARTIN, Engineer. 8vo. 50

MILES.—A PLAIN TREATISE ON HORSE-SHOEING.
With illustrations. By WILLIAM MILES, author of "The Horse's Foot," $1 00

MOLESWORTH. POCKET-BOOK OF USEFUL FORMULÆ AND MEMORANDA FOR CIVIL AND MECHANICAL ENGINEERS.
By GUILFORD L. MOLESWORTH, Member of the Institution of Civil Engineers, Chief Resident Engineer of the Ceylon Railway. Second American, from the Tenth London Edition. In one volume, full bound in pocket-book form . . $2 00

MOORE.—THE INVENTOR'S GUIDE:
Patent Office and Patent Laws; or, a Guide to Inventors, and a Book of Reference for Judges, Lawyers, Magistrates, and others. By J. G. MOORE. 12mo., cloth . . $1 25

MOREAU.—PRACTICAL GUIDE FOR THE JEWELLER,
In the Application of Harmony of Colors in the Arrangement of Precious Stones, Gold, etc., from the French of M. L. MOREAU, Jeweller and Designer. Illustrated. (In press.)

NAPIER.—CHEMISTRY APPLIED TO DYEING.
By JAMES NAPIER, F. C. S. A new and revised edition, brought down to the present condition of the Art. Illustrated. (In press.)

NAPIER.—A MANUAL OF DYEING RECEIPTS FOR GENERAL USE.
By JAMES NAPIER, F. C S. *With Numerous Patterns of Dyed Cloth and Silk.* Second edition, revised and enlarged. 12mo. $3 75

NAPIER.—MANUAL OF ELECTRO-METALLURGY:
Including the Application of the Art to Manufacturing Processes. By JAMES NAPIER. Fourth American, from the Fourth London edition, revised and enlarged. Illustrated by engravings. In one volume, 8vo. $2 00

NEWBERY.—GLEANINGS FROM ORNAMENTAL ART OF EVERY STYLE;
Drawn from Examples in the British, South Kensington, Indian, Crystal Palace, and other Museums, the Exhibitions of 1851 and 1862, and the best English and Foreign works. In a series of one hundred exquisitely drawn Plates, containing many hundred examples. By ROBERT NEWBERY. 4to. $15 00

NICHOLSON.—A MANUAL OF THE ART OF BOOK-BINDING:
Containing full instructions in the different Branches of Forwarding, Gilding, and Finishing. Also, the Art of Marbling Book-edges and Paper. By JAMES B. NICHOLSON. Illustrated. 12mo., cloth $2 25

NORRIS.—A HAND-BOOK FOR LOCOMOTIVE ENGINEERS AND MACHINISTS:
Comprising the Proportions and Calculations for Constructing Locomotives; Manner of Setting Valves; Tables of Squares, Cubes, Areas, etc. etc. By SEPTIMUS NORRIS, Civil and Mechanical Engineer. New edition. Illustrated, 12mo., cloth $2 00

NYSTROM.— ON TECHNOLOGICAL EDUCATION AND THE CONSTRUCTION OF SHIPS AND SCREW PROPELLERS:
For Naval and Marine Engineers. By JOHN W. NYSTROM, late Acting Chief Engineer U. S. N. Second edition, revised with additional matter. Illustrated by seven engravings. 12mo.
$2 50

O'NEILL.—CHEMISTRY OF CALICO PRINTING, DYEING, AND BLEACHING:
Including Silken, Woollen, and Mixed Goods; Practical and Theoretical. By CHARLES O'NEILL. (In press.)

O'NEILL.—A DICTIONARY OF CALICO PRINTING AND DYEING:
Containing a Brief Account of all the Substances and Processes in Use in the Arts of Printing and Dyeing Textile Fabrics; with Practical Receipts and Scientific Information. By CHARLES O'NEILL, Analytical Chemist, Fellow of the Chemical Society of London, etc. etc. Author of "Chemistry of Calico Printing and Dyeing." 8vo. (In press.)

OVERMAN—OSBORN.—THE MANUFACTURE OF IRON IN ALL ITS BRANCHES:
Including a Practical Description of the various Fuels and their Values, the Nature, Determination and Preparation of the Ore, the Erection and Management of Blast and other Furnaces, the characteristic results of Working by Charcoal, Coke, or Anthracite, the Conversion of the Crude into the various kinds of Wrought Iron, and the Methods adapted to this end. Also, a Description of Forge Hammers, Rolling Mills, Blast Engines, &c. &c. To which is added an Essay on the Manufacture of Steel. By FREDERICK OVERMAN, Mining Engineer. The whole thoroughly revised and enlarged, adapted to the latest Improvements and Discoveries, and the particular type of American Methods of Manufacture. With various new engravings illustrating the whole subject. By H. S. OSBORN, LL. D. Professor of Mining and Metallurgy in Lafayette College. In one volume, 8vo. (In press.) . $10 00

PAINTER, GILDER, AND VARNISHER'S COMPANION:
Containing Rules and Regulations in everything relating to the Arts of Painting, Gilding, Varnishing, and Glass Staining, with numerous useful and valuable Receipts; Tests for the Detection of Adulterations in Oils and Colors, and a statement of the Diseases and Accidents to which Painters, Gilders, and

Varnishers are particularly liable, with the simplest methods of Prevention and Remedy. With Directions for Graining. Marbling, Sign Writing, and Gilding on Glass. To which are added COMPLETE INSTRUCTIONS FOR COACH PAINTING AND VARNISHING. 12mo., cloth $1 50

PALLETT.—THE MILLER'S, MILLWRIGHT'S, AND ENGINEER'S GUIDE.
By HENRY PALLETT. Illustrated. In one vol. 12mo. $3 00

PERKINS.—GAS AND VENTILATION.
Practical Treatise on Gas and Ventilation. With Special Relation to Illuminating, Heating, and Cooking by Gas. Including Scientific Helps to Engineer-students and others. With illustrated Diagrams. By E. E. PERKINS. 12mo., cloth $1 25

PERKINS AND STOWE.—A NEW GUIDE TO THE SHEET-IRON AND BOILER PLATE ROLLER:
Containing a Series of Tables showing the Weight of Slabs and Piles to Produce Boiler Plates, and of the Weight of Piles and the Sizes of Bars to produce Sheet-iron; the Thickness of the Bar Gauge in Decimals; the Weight per foot, and the Thickness on the Bar or Wire Gauge of the fractional parts of an inch; the Weight per sheet, and the Thickness on the Wire Gauge of Sheet-iron of various dimensions to weigh 112 lbs. per bundle; and the conversion of Short Weight into Long Weight, and Long Weight into Short. Estimated and collected by G. H. PERKINS and J. G. STOWE $2 50

PHILLIPS AND DARLINGTON.—RECORDS OF MINING AND METALLURGY:
Or Facts and Memoranda for the use of the Mine Agent and Smelter. By J. ARTHUR PHILLIPS, Mining Engineer, Graduate of the Imperial School of Mines, France, etc., and JOHN DARLINGTON. Illustrated by numerous engravings. In one volume, 12mo. $2 00

PRADAL, MALEPEYRE, AND DUSSAUCE. — A COMPLETE TREATISE ON PERFUMERY:
Containing notices of the Raw Material used in the Art, and the Best Formulæ. According to the most approved Methods followed in France, England, and the United States. By M. P. PRADAL, Perfumer Chemist, and M. F. MALEPEYRE. Translated from the French, with extensive additions, by Professor H. DUSSAUCE. 8vo. $10 00

PROTEAUX.—PRACTICAL GUIDE FOR THE MANUFACTURE OF PAPER AND BOARDS.

By A. Proteaux, Civil Engineer, and Graduate of the School of Arts and Manufactures, Director of Thiers's Paper Mill, 'Puy-de-Dôme. With additions, by L. S. Le Normand. Translated from the French, with Notes, by Horatio Paine, A. B., M. D. To which is added a Chapter on the Manufacture of Paper from Wood in the United States, by Henry T. Brown, of the "American Artisan." Illustrated by six plates, containing Drawings of Raw Materials, Machinery, Plans of Paper-Mills, etc. etc. 8vo. $5 00

REGNAULT.—ELEMENTS OF CHEMISTRY.

By M. V. Regnault. Translated from the French by T. Forrest Betton, M. D., and edited, with notes, by James C. Booth, Melter and Refiner U. S. Mint, and Wm. L. Faber, Metallurgist and Mining Engineer. Illustrated by nearly 700 wood engravings. Comprising nearly 1500 pages. In two volumes, 8vo., cloth $10 00

SELLERS.—THE COLOR MIXER:

Containing nearly Four Hundred Receipts for Colors, Pastes, Acids, Pulps, Blue Vats, Liquors, etc. etc., for Cotton and Woollen Goods: including the celebrated Barrow Delaine Colors. By John Sellers, an experienced Practical Workman. In one volume, 12mo. $2 50

SHUNK—A PRACTICAL TREATISE ON RAILWAY CURVES AND LOCATION, FOR YOUNG ENGINEERS.

By Wm. F. Shunk, Civil Engineer. 12mo. . . $1 50

SMEATON—BUILDER'S POCKET COMPANION:

Containing the Elements of Building, Surveying, and Architecture; with Practical Rules and Instructions connected with the subject. By A. C. Smeaton, Civil Engineer, etc. In one volume, 12mo. $1 25

SMITH—THE DYER'S INSTRUCTOR:

Comprising Practical Instructions in the Art of Dyeing Silk, Cotton, Wool, and Worsted, and Woollen Goods: containing nearly 800 Receipts. To which is added a Treatise on the Art of Padding; and the Printing of Silk Warps, Skeins, and Handkerchiefs, and the various Mordants and Colors for the different styles of such work. By David Smith, Pattern Dyer. 12mo., cloth. $3 00

SMITH.—PARKS AND PLEASURE GROUNDS:
Or Practical Notes on Country Residences, Villas, Public Parks, and Gardens. By CHARLES H. J. SMITH, Landscape Gardener and Garden Architect, etc. etc. 12mo. . $2 25

STOKES.—CABINET-MAKER'S AND UPHOLSTERER'S COMPANION:
Comprising the Rudiments and Principles of Cabinet-making and Upholstery, with Familiar Instructions, Illustrated by Examples for attaining a Proficiency in the Art of Drawing, as applicable to Cabinet-work; The Processes of Veneering, Inlaying, and Buhl-work; the Art of Dyeing and Staining Wood, Bone, Tortoise Shell, etc. Directions for Lackering, Japanning, and Varnishing; to make French Polish; to prepare the Best Glues, Cements, and Compositions, and a number of Receipts particularly for workmen generally. By J. STOKES. In one vol. 12mo. With illustrations $1 25

STRENGTH AND OTHER PROPERTIES OF METALS.
Reports of Experiments on the Strength and other Properties of Metals for Cannon. With a Description of the Machines for Testing Metals, and of the Classification of Cannon in service. By Officers of the Ordnance Department U. S. Army. By authority of the Secretary of War. Illustrated by 25 large steel plates. In 1 vol. quarto $10 00

TABLES SHOWING THE WEIGHT OF ROUND, SQUARE, AND FLAT BAR IRON,-STEEL, ETC.,
By Measurement. Cloth 63

TAYLOR.—STATISTICS OF COAL:
Including Mineral Bituminous Substances employed in Arts and Manufactures; with their Geographical, Geological, and Commercial Distribution and amount of Production and Consumption on the American Continent. With Incidental Statistics of the Iron Manufacture. By R. C. TAYLOR. Second edition, revised by S. S. HALDEMAN. Illustrated by five Maps and many wood engravings. 8vo., cloth . . . $3 00

TEMPLETON.—THE PRACTICAL EXAMINATOR ON STEAM AND THE STEAM-ENGINE:
With Instructive References relative thereto, for the Use of Engineers, Students, and others. By WM. TEMPLETON, Engineer. 12mo. $1 25

THOMAS.—THE MODERN PRACTICE OF PHOTOGRAPHY.
By R. W. Thomas, F. C. S. 8vo., cloth . . . 75

THOMSON.—FREIGHT CHARGES CALCULATOR.
By Andrew Thomson, Freight Agent . . . $1 25

TURNBULL.—THE ELECTRO-MAGNETIC TELEGRAPH:
With an Historical Account of its Rise, Progress, and Present Condition. Also, Practical Suggestions in regard to Insulation and Protection from the effects of Lightning. Together with an Appendix, containing several important Telegraphic Devices and Laws. By Lawrence Turnbull, M. D., Lecturer on Technical Chemistry at the Franklin Institute. Revised and improved. Illustrated. 8vo. . . . $3 00

TURNER'S (THE) COMPANION:
Containing Instructions in Concentric, Elliptic, and Eccentric Turning; also various Plates of Chucks, Tools, and Instruments; and Directions for using the Eccentric Cutter, Drill, Vertical Cutter, and Circular Rest; with Patterns and Instructions for working them. A new edition in one vol. 12mo.
$1 50

ULRICH—DUSSAUCE.—A COMPLETE TREATISE ON THE ART OF DYEING COTTON AND WOOL:
As practised in Paris, Rouen, Mulhausen, and Germany. From the French of M. Louis Ulrich, a Practical Dyer in the principal Manufactories of Paris, Rouen, Mulhausen, etc. etc.; to which are added the most important Receipts for Dyeing Wool, as practised in the Manufacture Impériale des Gobelins, Paris. By Professor H. Dussauce. 12mo. $3 00

URBIN—BRULL.—A PRACTICAL GUIDE FOR PUDDLING IRON AND STEEL.
By Ed. Urbin, Engineer of Arts and Manufactures. A Prize Essay read before the Association of Engineers, Graduate of the School of Mines, of Liege, Belgium, at the Meeting of 1865—6. To which is added a Comparison of the Resisting Properties of Iron and Steel. By A. Brull. Translated from the French by A. A. Fesquet, Chemist and Engineer. In one volume, 8vo. $1 00

WATSON.—A MANUAL OF THE HAND-LATHE.
By Egbert P. Watson, Late of the "Scientific American," Author of "Modern Practice of American Machinists and Engineers." In one volume, 12mo. (In press.)

WATSON.—THE MODERN PRACTICE OF AMERICAN MACHINISTS AND ENGINEERS:
Including the Construction, Application, and Use of Drills, Lathe Tools, Cutters for Boring Cylinders, and Hollow Work Generally, with the most Economical Speed of the same, the Results verified by Actual Practice at the Lathe, the Vice, and on the Floor. Together with Workshop management, Economy of Manufacture, the Steam-Engine, Boilers, Gears, Belting, etc. etc. By EGBERT P. WATSON, late of the "Scientific American." Illustrated by eighty-six engravings. 12mo. . . . $2 50

WATSON.—THE THEORY AND PRACTICE OF THE ART OF WEAVING BY HAND AND POWER:
With Calculations and Tables for the use of those connected with the Trade. By JOHN WATSON, Manufacturer and Practical Machine Maker. Illustrated by large drawings of the best Power-Looms. 8vo. $7 50

WEATHERLY.—TREATISE ON THE ART OF BOILING SUGAR, CRYSTALLIZING, LOZENGE-MAKING, COMFITS, GUM GOODS,
And other processes for Confectionery, &c. In which are explained, in an easy and familiar manner, the various Methods of Manufacturing every description of Raw and Refined sugar Goods, as sold by Confectioners and others . . $2 00

WILL.—TABLES FOR QUALITATIVE CHEMICAL ANALYSIS.
By Prof. HEINRICH WILL, of Giessen, Germany. Seventh edition. Translated by CHARLES F. HIMES, Ph. D., Professor of Natural Science, Dickinson College, Carlisle, Pa. . $1 25

WILLIAMS.—ON HEAT AND STEAM:
Embracing New Views of Vaporization, Condensation, and Expansion. By CHARLES WYE WILLIAMS, A. I. C. E. Illustrated. 8vo. $3 50

www.ingramcontent.com/pod-product-compliance
Lightning Source LLC
Chambersburg PA
CBHW032102230426
43672CB00009B/1617